THE WORLD'S SAVANNAS

Economic Driving Forces, Ecological Constraints and
Policy Options for Sustainable Land Use

MAN AND THE BIOSPHERE SERIES

MAN AND THE BIOSPHERE SERIES

Series Editor J.N.R. Jeffers

VOLUME 12

THE WORLD'S SAVANNAS
Economic Driving Forces, Ecological Constraints and Policy Options for Sustainable Land Use

Edited by
M.D. Young and O.T. Solbrig

PUBLISHED BY

PARIS

AND

The Parthenon Publishing Group
International Publishers in Science, Technology & Education

Published in 1993 by the United Nations Educational, Scientific and Cultural Organization,
7 Place de Fontenoy, 75700 Paris, France—UNESCO ISBN 92-3-102750-6

and

The Parthenon Publishing Group Limited
Casterton Hall, Carnforth,
Lancs LA6 2LA, UK—ISBN 1-85070-417-1

and

The Parthenon Publishing Group Inc.
One Blue Hill Plaza
PO Box 1564, Pearl River,
New York 10965, USA—ISBN 1-85070-417-1

Typeset by Lasertext Ltd, Stretford, Manchester
Printed and bound in Great Britain by
Butler and Tanner Ltd., Frome and London

British Library Cataloguing in Publication Data

World's Savannas: Economic Driving Forces, Ecological Constraints and Policy
Options for Sustainable Land Use. – (Man & the Biosphere Series; Vol. 12)
I. Young, M. D. II. Solbrig, O. T. III. Series
333.74

ISBN 1-85070-417-1

Library of Congress Cataloging-in-Publication Data

The World's savannas: economic driving forces, ecological constraints
and policy options for sustainable land use/edited by M.D. Young
and O.T. Solbrig.
    p.   cm. — (Man and the biosphere series: v. 12)
    Includes bibliographical references and index.
    ISBN 1-85070-417-1
    1. Land use — Environmental aspects — Congresses.
2. Savannas — Congresses. 3. Conservation of natural resources
— Congresses. 4. Land tenure — Congresses. 5. Sustainable
agriculture — Congresses. 6. Land use — Government policy
— Congresses. I. Young. M.D. (Michael Denis), 1952– II. Solbrig,
Otto Thomas. III. Series.
HD105.W675 1993
333.76 — dc20   92-41899
                                                                      CIP

PREFACE

UNESCO's Man and the Biosphere Programme

Improving scientific understanding of natural and social processes relating to man's interactions with his environment, providing information useful to decision-making on resource use, promoting the conservation of genetic diversity as an integral part of land management, enjoining the efforts of scientists, policy-makers and local people in problem-solving ventures, mobilizing resources for field activities, strengthening of regional co-operative frameworks, these are some of the generic characteristics of UNESCO's Man and the Biosphere (MAB) Programme.

The MAB Programme was launched in the early 1970s. It is a nationally based, international programme of research, training, demonstration and information diffusion. The overall aim is to contribute to efforts for providing the scientific basis and trained personnel needed to deal with problems of rational utilization and conservation of resources and resource systems, and problems of human settlements. MAB emphasizes research for solving problems: it thus involves research by interdisciplinary teams on the interactions between ecological and social systems; field training; and applying a systems approach to understanding the relationships between the natural and human components of development and environmental management.

MAB is a decentralized programme with field projects and training activities in all regions of the world. These are carried out by scientists and technicians from universities, academies of sciences, national research laboratories and other research and development institutions, under the auspices of more than a hundred MAB National Committees. Activities are undertaken in co-operation with a range of international governmental and non-governmental organizations.

Man and the Biosphere Book Series

The Man and the Biosphere Series was launched with the aim of communicating some of the results generated by the MAB Programme and is aimed primarily at upper level university students, scientists and resource managers, who are not necessarily specialists in ecology. The books are not normally suitable for undergraduate text books but rather provide additional resource material in the form of case studies based on primary data collection and written by the researchers involved; global and regional syntheses of comparative research conducted in several sites or countries; and state-of-the-art assessments of knowledge or methodological approaches based on scientific meetings, commissioned reports or panels of experts. The Series Editor is John Jeffers, formerly Director of the Institute of Terrestrial Ecology, in the United Kingdom, who has been associated with MAB since its inception.

The World's Savannas

The present volume addresses the issue of managing tropical savannas in ways that are more environmentally sustainable, economically profitable and socially equitable than at present.

The volume represents one output from an international workshop held at UNEP Headquarters in Nairobi (Kenya) from 23–26 January 1991, around the theme of *Economic Driving Forces and Ecological Constraints in Savanna Land Use.* A joint initiative of the International Union of Biological Sciences (IUBS), UNESCO, the United Nations Environment Programme (UNEP) and the Commission of European Communities, the workshop was organized as one of several synthesis activities of the collaborative IUBS–MAB programme on Responses of Savannas to Stress and Disturbance (RSSD). The workshop also related to activities within the MAB research orientation on 'Human investment and resource use' with its focus on issues at the interface of the policy goals of Environmental integrity, Economic efficiency and Equity.

The principal aims of the Nairobi workshop were to assess and, as far as possible, quantify the effects of agricultural policies on the development and sustained use of savannas; to identify ecological, social and economic constraints to savanna development; and to identify the nature of policy changes necessary to enhance prospects for economic growth and development throughout the savannas. The emphasis was on the development of policy recommendations at a national and international level. Some twenty invited specialists took part, with backgrounds in ecology, economics and related subjects. The workshop was based on a number of issue papers and case studies. Issue papers addressed such topics as ecological constraints to savanna land use, modifying land tenure arrangements to improve savanna land use, and the effects of international trade arrangements and the policies of other countries on savanna land use. Case studies included accounts of savanna land use in such countries as Australia, Brazil, Kenya, South Africa, Tanzania, Venezuela and Zimbabwe. Most case studies were prepared jointly by a social scientist and an ecologist.

Three main types of product have been prepared by the two-person team who were principally responsible for the workshop concept, design and implementation, resource economist Michael D. Young and biologist Otto T. Solbrig: overview articles in popular environmental magazines, such as that in the April 1992 issue of *Environment*; a 50-page synoptic report published by UNESCO early in 1993 as MAB Digest 12; and the present multi-authored volume. It is hoped that this book will be of interest to those who study the ecology and societies of tropical savannas as well as those who shape, move and implement policies which affect savanna land use.

MAN AND THE BIOSPHERE SERIES

To Malcolm Hadley
and
Talal Younès

They have done much
for science
throughout the world

CONTENTS

D.H.M. Cumming
WWF Multispecies Animal
  Production Systems Project
PO Box 8437
Causeway
Harare
Zimbabwe

M. Gadgil
Centre for Ecological Sciences and
  Jawaharlal Nehru Centre for
  Advanced Scientific Research
Indian Institute of Science
Bangalore 560012
India

J.H. Holmes
Department of Geographical
  Sciences
University of Queensland
St. Lucia
Queensland 4072
Australia

M. Kituyi
African Centre for Technology
  Studies
PO Box 45917
Nairobi
Kenya

C.A. Klink
Department of Organismic and
  Evolutionary Biology
Harvard University Herbaria
22 Divinity Avenue
Cambridge
Massachusetts 02138
USA

C. Lane
Drylands Programme
International Institute for
  Environment and Development
3 Endsleigh Street
London WC1H 0DD
UK

M.T. Mentis
Envirobiz Africa
77 Tanner Road
Wembley
Pietermaritzburg 3201
South Africa

A.G. Moreira
Department of Organismic and
  Evolutionary Biology
Harvard University Herbaria
22 Divinity Avenue
Cambridge
Massachusetts 02138
USA

A. Moreno
Depto. de Antropología Sociología
Facultad de Humanidades y
  Educación
Universidad de los Andes
Mérida
Venezuela

J.J. Mott
Department of Agriculture
University of Queensland
St. Lucia
Queensland 4072
Australia

M.W. Murphree
Centre for Applied Social Sciences
University of Zimbabwe
PO Box MP167
Mount Pleasant
Harare
Zimbabwe

D.W. Pearce
Centre for Social and Economic
  Research on the Global
  Environment
University College London
Gower Street
London WC1E 6BT
UK

I. Scoones
Drylands Programme
International Institute for
  Environment and Development
3 Endsleigh Street
London WC1H 0DD
UK

N. Seijas
Free Market Foundation of
  Southern Africa
PO Box 52713
Saxonwold 2132
South Africa

J.F. Silva
Centro de Investigaciones
  Ecológicas de los Andes
  Tropicales (CIELAT)
Facultad de Ciencas
Universidad de los Andes
Mérida
Venezuela

O.T. Solbrig
Department of Organismic and
  Evolutionary Biology
Harvard University
22 Divinity Avenue
Cambridge
Massachusetts 02138
USA

C. Toulmin
Drylands Programme
International Institute for
  Environment and Development
3 Endsleigh Street
London WC1H 0DD
UK

M.D. Young
CSIRO Division of Wildlife and
  Ecology
PO Box 84
Lyneham
Canberra
Australia 2062

FOREWORD

Made possible by a generous grant from the Commission of the European Communities, this book focuses on the world's savanna lands. Unlike previous work in this area, it makes a serious attempt to bring together the social and physical sciences in an integrated framework.

The initial proposal, which led to this ambitious project, came from the scientists participating in the International Union of Biological Sciences (IUBS) Programme on the "Decade of the Tropics". Seeking an opportunity to interact and collaborate with social scientists, they proposed a series of collaborative studies that would focus on the interface between the two disciplines. Their goal was a set of holistic policy recommendations that would be neither ecologically, nor socially or economically naive. The UNESCO Man and the Biosphere Programme focus on "Human Investment and Resource Use" which centred on the "3 Es" – economic efficiency, equity and environmental integrity – provided an excellent framework to begin discussions.

The format we took to encourage collaboration was to invite teams, usually of an ecologist and a social scientist, to prepare a case study on a part of the world's savannas and bring this to a workshop in Nairobi. Organising that workshop involved a large amount of work and its success was due to the efforts of Talal Younès and Colleen Adam at IUBS in Paris and Malcolm Hadley of UNESCO, Paris. Local arrangements were organised excellently by Ali Ayoub at United Nations Environment Programme (UNEP) in Nairobi. The willingness with which UNEP made its conference facilities avaliable to us in Nairobi is gratefully acknowledged.

Unfortunately, it was not possible for the two people who prepared the South African case study – Mike Mentis and Nancy Seijas – to attend the workshop. Their important contribution, however, is gratefully acknowledged and included in this book. We are also grateful to Madhav Gadgil who, following the workshop, prepared an Indian case study for us.

xix

Similarly, we would like to thank David Pearce for permission to reproduce a case study that he had prepared on Botswana's savannas. We would also like to thank Jimoh Omo-Fadaka and Mike Norton-Griffiths for the significant contributions that they made to the workshop.

Following the workshop, participants revised their papers in the usual manner. As we all know, however, turning any set of manuscripts into a book involves a lot of typing and editorial work. Recognising this we would particularly like to acknowledge Kate Ransley's word processing skills and Raylee Singh's editorial skills. We are deeply indebted to both of them.

M.D. Young
and
O.T. Solbrig

Section 1

Introduction

CHAPTER 1

ECONOMIC AND ECOLOGICAL DRIVING FORCES
AFFECTING TROPICAL SAVANNAS

O.T. Solbrig and M.D. Young

INTRODUCTION

According to Bourlière and Hadley (1983) the term "savanna" designates
a tropical grassland with scattered trees. Defined this way, savannas are
the most common tropical landscape unit. They are found in all four
tropical continents and in more than 20 countries (see Figure 1.1).

Savanna vegetation results from the tropical monsoon pattern of rainfall
with a yearly alternation of a rainy and a dry season. In areas of high
rainfall and no or short dry seasons, trees dominate and rainforests result.
In areas of very low rainfall and very short wet seasons, we find deserts
and semi-desert shrublands. Between these extremes of climate, we find a
mixed vegetation of trees and grasses, called savannas.

Given the broad definition of what constitutes a savanna, it is not
surprising that there is a great deal of variability in most of their physical
and biological characteristics: amount of rainfall, length of the dry season,
soil characteristics, and density of trees, fauna, and micro-organisms. This
variability, in turn, determines the types of human occupation and use.
Historical factors relating to the diffusion of the human race throughout
the globe since its African origins determine the time humans have occupied
savannas. People have lived in African savannas for more than a million
years, but in the American savannas for less than 30,000 years.

Situated between rainforests and deserts, savannas have a comparative
advantage in tourism, meat and fibre production and, to a lesser extent,
crop production. They are a classic example of a conditionally renewable
resource. Until recently savannas were not used intensively. In the last 50
years, increasing human populations in the tropics have intensified the use
of these ecosystems, to the point that it is feared that they could become

3

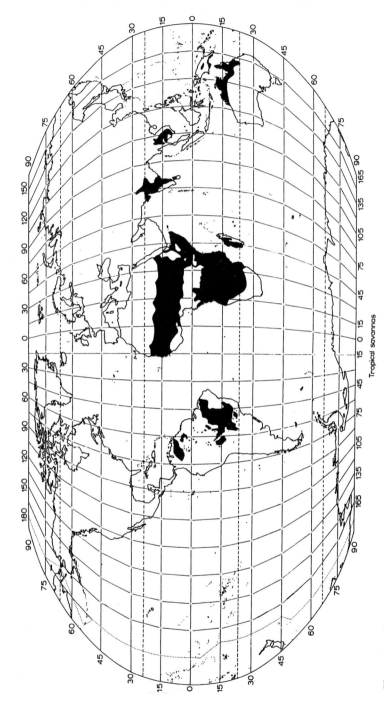

Figure 1.1 Distribution of the world's savannas

Tropical savannas

seriously degraded, especially the dry savannas (i.e. those with less than 500–600 mm yearly rainfall).

Savannas found in the southern part of the Sahara were a focus of global attention during the 1970s, but the fate, status and needs of all the world's savannas have never been examined within an integrated framework which combines economic, social and environmental considerations. This book attempts to rectify this deficiency. It arises from the collective efforts of a group of ecologists and social scientists who met in Nairobi, Kenya in January 1991 to review a series of general policy and regional case studies they had prepared on the above topic. The book focuses on the conditions necessary to ensure that savanna resources remain of value to future generations whilst contributing to the needs of those who aspire to use them today.

In this introductory chapter some of the general issues regarding savanna use and economic policies are reviewed. The objective is to call attention to the principal problems affecting the development of tropical savannas. More details are provided in the issue chapters and case studies that follow. First we provide some general background, including a brief discussion of environmental determinism. We then define environment, resources and ecosystem, and some theoretical concepts regarding their meaning. We follow by presenting a graphical model of land use, and conclude with a presentation of the major policy issues regarding development and land tenure in savannas. The last chapter presents a synthesis of the policy proposals and recommendations that emerged from the Nairobi meeting.

BACKGROUND

These days, few ecological topics evoke more emotional reaction among concerned citizens of developed countries than the prospect of a drastic transformation of natural landscapes. Some maintain that the opening of new areas to agriculture, especially in the tropics, is the forerunner of acute deleterious changes in climate and ecosystem stability. Many express increasing alarm about the high rate of deforestation in the tropics, fearing that it will lead to loss of species and changes in rainfall patterns and an increase in atmospheric carbon dioxide. There is also the fear that the Earth will experience great global climatic changes in the next 50–100 years, in response to the increase in atmospheric greenhouse gases caused by human activities.

Often, among governments and inhabitants of tropical countries, there is more interest in the development of savannas and other vegetation types than in their conservation. Governments and the public in these countries perceive development – for agriculture, livestock raising, and mining – of unused or sparsely used regions as a solution to problems brought about by high population growth and low incomes. Savannas figure prominently among areas perceived as unused.

The different viewpoints regarding the future use of tropical landscapes have created a conflict between those who favour development and those who support conservation. This conflict needs to be resolved, and can be resolved. There is, of course, no way of developing an area without creating some ecological disruption. However, better understanding of the function of the ecosystem, and appropriate management and investment, can keep such disruptions from becoming a serious environmental threat. Human alteration of natural landscapes is not new. People began to transform their surroundings with the adoption of agriculture and animal domestication some 10,000 years ago. The invention of the plough and the diffusion of plough-based agriculture accelerated environmental modification (Norgaard 1987; Solbrig 1991). The process intensified again with the industrial revolution and has reached its apogee in the last 20 years with the introduction of high-input farming techniques known as the 'Green Revolution'.

Environmental determinism

The concern about landscape transformation arises in part from the public notion that the characteristics of the environment are fixed and more or less immutable, and that they constrain, if not determine, human activity. These views are sometimes also assumed by scholars. Natural scientists in particular are prone to hold these ideas. But even among social scientists there was a school of environmental determinism which is now largely discredited (Ratzel 1882; Semple 1911; Meggars 1954, 1957). All the papers in this book suggest that there are few environmental limits that cannot be overcome through human effort and ingenuity. We believe that the environment is only one of many constraints affecting human behaviour. Like other constraints, environmental barriers are not absolute and most can be overcome by investing money, energy and/or materials. It is these costs that determine the limits to human activities, not some magical law of nature.

The concept of environmental determinism belongs to a long tradition that considers the natural world separately from the human sphere; that regards people as observers looking into nature and beholden to it, rather than as integral interacting constituents of it. For example, people are deemed by all big religions to be the product of some special creation and the object of special attention by the deity, often to protect them from the determinism of their surroundings. Occasionally the scientific approach to

6

the study of nature also reflects this attitude which arose with the development of agriculture. Hunter-gatherers have a much greater regard for nature than agriculturalists (Oelschlaeger 1991). We maintain that human behaviour is as "natural" as that of a bacterium or an earthworm. In the words of Sober (1986, his italics):

> *...If we are part of nature, then everything we do is part of nature, and is natural in that primary sense.* When we domesticate organisms and bring them into a state of dependence on us, this is simply an example of one species exerting a selection pressure on another. If one calls this "unnatural", one might just as well say the same of parasitism or symbiosis...

The physical and biological components of ecosystems (including people) interact in complex and intricate ways. Every species, including *Homo sapiens*, adjusts its characteristics – behavioural, demographic and genetic – to try to maximise its fitness. The structural and functional characteristics of the ecosystem are the result of the interactions of all these forces. But all species, including our own, act with incomplete information regarding the behaviour of others. Furthermore, in order to evolve and adapt to their surroundings, all species, but especially those without a well-developed sensory system, must rely on the appearance through mutation and recombination of suitable genetic information.

Land use

The ways by which human societies extract energy and resources from the landscape is called land use. Land use reflects the characteristics – physical, behavioural and social – of a human society. It is a reflection of the level of technological development of the society, of its needs and of its values. In brief it is an accurate reflection of how a given society adapts to the environment.

Savanna land use, including what now is called, though not totally accurately, "sustainable development", is principally a function of technology and investment. By this we mean investment of money, energy and knowledge, and the use of appropriate procedures. Changes in technology or investment brought about predominantly by social evolution, or occasionally by organic evolution, cause changes in patterns of land use. These changes are part of the constant process of adaptation to the environment that human societies (and other organisms) undergo. This does not mean that societies do not have to contend with environmental constraints, but that constraints are not absolute and not confined to the outside environment. Rather, constraints are a function of the way human

7

societies are organised and behave, and how the rest of the ecosystem reacts to that behaviour. Ecological constraints can only be understood and studied in combination with the human societies that live in an area.

ENVIRONMENT, RESOURCES AND ECOSYSTEM

Environment, resources and ecosystem are defined in this section, since the significant differences between them are not always stated clearly, creating some problems in interpretation.

The environment is the sum total of the biotic and abiotic elements that can potentially interact with a biological system (this system can be one or many cells, individuals, populations, species or communities). It includes all that is external to and potentially influential upon an object of investigation. Usually the environment is defined in relation to an individual organism, but communities, societies and biotas also have environments. The environment is defined in terms of each separate object. So, for example, the environment of a biological community is different from that of the organisms within it. The biotic elements that constitute the environment of an organism are the relationships with other organisms, such as competition for resources, mutualisms, predation, etc. The abiotic elements of the environment are the physical and chemical states and characteristics of the surroundings, such as air temperature, amount of precipitation or soil texture. The term "environment" denotes an open-ended concept.

Resources are the specific physical components of our surroundings that an organism needs or uses to sustain itself, such as minerals, water or air. Some plants and animals are also resources for other species. Characteristics of the environment are not resources. So, for example, many plants and animals are resources, but biodiversity (a property of the biota) is part of the environment but is not a resource. A resource is defined in terms of a user. It denotes not a property but a relationship.

Organisms do not exist in physical and biological isolation. There is an interplay both among different species populations and among species populations and the physical and chemical components of the environment. An ecosystem is the sum total of organisms, their resources and their environment. It is an arrangement of mutual dependences by which the whole operates as a unit. Since all individuals and populations including humans exist and operate as parts of ecosystems, the basic unit of ecology is the ecosystem.

The concept of the ecosystem is often abused. In popular writings it has become equated with the sum of the geology, plants, microorganisms and animals of a particular region. From the perspective of an ecologist, the living world is an elaborate conglomerate of ecological units within

8

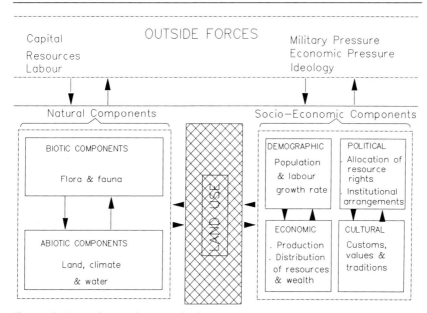

Figure 1.2 Determinants of savanna land use

ecological units, some discrete and tangible, others indistinct but no less real (O'Neil et al. 1986). Organisms are formed of organs, tissues, cells, and subcellular organelles. Each performs a specific function, from assembling chemical molecules into new cells and organisms, to capturing food and energy from its environment. But each organism interacts with others – sometimes competing for food, mates or territory, sometimes collaborating in myriads of ways. Furthermore, every organism is potentially food for another, from viruses and bacteria to the biggest of vertebrates, including people. In an ultimate sense, all organisms on the globe are part of a hierarchically organised network of networks. The whole is "the" global ecosystem, but each unit down to the cell and below is also an ecosystem. The key relation is not place, but connectedness.

ECOSYSTEMS, PEOPLE AND LAND USE

Figure 1.2 is a depiction of a model of an ecosystem that includes people and emphasises the relation between humans and their surroundings. On the left are the elements that natural scientists normally emphasise in their studies: climate, geology, flora and fauna, and their interrelationships. On the right are the elements usually studied by social scientists: the demographic,

9

economic, cultural and political components of human societies, and their interrelationships. In between is a box labelled land use.

This model in the broadest sense represents the way in which societies obtain from their surroundings the resources needed for their sustenance and growth. To obtain energy and materials from the environment (indicated by a right-pointing arrow between land and society) humans use technology, and invest effort and knowledge (indicated by a left-pointing arrow between society and land). This in turn modifies the "natural" components of the ecosystem (indicated by a left arrow between land and "nature"), which results in modifications to the system (indicated by a right arrow between "nature" and land), which ultimately controls what humans reap from the land.

The point that is emphasised in the figure is that the ecosystem is a dynamic system, and that in principle (but not necessarily in practice) most changes in the human-land relation caused by people are reversible. If the land-use technologies applied to obtain resources from the environment result in decreased outputs, people can modify their techniques. For this to happen, human societies have to have the organisation, the knowledge, the resources, and the institutions to deal effectively with any problem that may arise. Because we are dealing with a system, there are also inevitable time lags. So, for example, the use of flood irrigation leads to soil salinisation. The process is a slow one, however, and 10s, even 100s, of years can elapse before the levels of soil salinisation became so high as to make the land unusable. By then, the costs of reversing the process are prohibitive and the land is abandoned. This is what happened in ancient Babylon. Thus, the negative consequences of human behaviour must be identified and dealt with in advance of their occurrence. To do so, the potential effects of new technologies must be thoroughly researched, which requires appropriate institutions and a suitable knowledge base. Land-use history is replete with instances where people did not anticipate the negative consequences of their actions, or were unable to take appropriate and timely action because of lack of knowledge, resources or organisation. Lack of human investment (of capital, labour and knowledge) in developing appropriate land-use practices is as great an ecological constraint to land use as are climate and soils.

ENVIRONMENTAL MODIFICATION

Land use provides the best indicator of how humans interact with and feel about their environment. The closer people feel to the world of plants and animals, to the forests and the mountains, to the fields and to the savannas, the more likely their land-use practices will be in harmony with their surroundings. On the other hand, if a human society sees its dependency

10

on the biota and the geology as remote or unimportant, then it is more likely that it will engage in exploitative land-use practices. Since the latter attitude is non-sustainable, eventually all human societies must find a way of life compatible with their environment, or they will not be able to survive. Industrial society, with its associated high degree of urbanisation, has detached people from nature and made them more insensitive to the negative consequences of their land-use practices.

One of the effects of human intervention in nature through land use is the simplification of plant and animal communities. This is inevitable: the flux of solar energy and the total amount of mineral ions on the Earth is fixed. If humans (or any other species) capture increasing proportions of solar energy and materials to sustain themselves and their civilisations, that energy and those materials are no longer available to maintain as many other organisms as before. The extreme case of the simplification of nature takes place when people replace a complex natural community, such as a tropical forest, with agricultural fields. Such simple agricultural systems are often considered to be more fragile, less resilient to environmental fluctuations, and less "sustainable" as a resource (Brown 1981; Fernside 1986). Yet simple, high-intensity agricultural systems are not necessarily inherently more fragile. It will depend on the land-use practices of the human society. Since they require a stream of inputs – fertilisers, water and, above all, knowledge – they are only viable in high-income and politically stable societies.

Forestry, livestock grazing, and especially agriculture reduce the diversity of natural communities of plants, animals and microorganisms. Normally, the more technified the society and its land-use practices, the greater the simplification they bring about. Yet, what is simplified in these instances is not the entire ecosystem, but only a part of it. The human component of the ecosystem is usually more intricate the greater the simplification people can induce on the "natural" side of the ecosystem (Hawley 1986). It is not known whether the total ecosystem (including the human components) increases or decreases in overall complexity because we lack a clear way of measuring complexity. Some characteristics, such as ecological productivity, can actually increase with human intervention when it results in higher rates of circulation of energy and materials.

Ecological dislocations often result from oversimplification of the environment by societies that fail to make appropriate investments in land use because of lack of knowledge, finance or the necessary organisation (Adámoli and Fernandez 1980; Foweraker 1981). Typical examples resulting from lack of foresight and planning are frontier societies such as those in the Amazonian forest (Fernside 1986). Ecosystems become degraded (i.e. their overall structure and function is simplified and their productivity reduced) when societies simplify the natural communities, extracting energy

11

and materials, without at the same time increasing the rate of investment by developing more appropriate technologies. Sustainable land-use practices result from a balance between the intensity of use and the complexity of the human institutions engaged in the use. Such groups include farmers, agricultural research establishments, market forces, resource rights and administrative structures.

Beliefs, rituals, and social and political interrelations in all human societies are intricate and serve the needs and aspirations of the members of each culture. Civilisations differ in their technological development, especially transportation and communication technology, in their overall knowledge about nature, and in their use of outside sources of energy. Considerable organisation and division of labour is required in order to use outside energy sources effectively, and to develop complex technologies. Therefore, we see a relation between the complexity of a human society and its capacity to simplify the natural part of the ecosystem in order to increase productivity without creating permanent degradation. This is illustrated by the diversity and effectiveness of the different uses of savanna lands.

POLICY ISSUES

According to some, the goal of a land-use policy is to maximise profits without incurring external costs, such as air or river pollution. This is not a realistic objective. The laws of physics tell us that no activity that involves work can take place without an increase in the entropy of the system. This includes living organisms. Living systems are open thermodynamic systems, constantly receiving energy inputs from the sun. They generate waste products in the form of heat and degraded materials, such as CO_2, O_2, CH_4, and other gases. The activities of organisms over geological eras has led to a dynamic equilibrium of these waste materials in the sun-earth-atmosphere system that we now regard as "natural".

Humans must obey the same physical rules that govern the behaviour of all other organisms. To change the "natural" state of the environment we must perform work. This generates heat and other by-products, which we call "pollutants". This is as true of traditional hunter-gatherers as it is of present-day technological societies. The difference between these two societies lies in the degree to which they can tap outside sources of energy to perform work. While the hunter-gatherer had only his own energy, people are now able to tap a variety of energy sources, from hydroelectric to solar. Today the principal outside source of energy is the burning of coal and petroleum. Since much more energy is going in per unit time, many more by-products per unit time are produced.

12

Other natural forces are at work too, particularly geological forces. Wind and water and gravity combine to level all surfaces on the face of the Earth, given enough time. Other forces, resulting from the movement of the continents, produce the exact opposite effect, namely the raising of mountains. The interaction of processes that give rise to mountains with erosional forces that level them results in the changing topography of the planet. Erosion is as natural in the history of the earth as the production of by-products by organisms. Humans influence the rates of erosion when they remove or alter vegetation cover, build roads or construct dams which change water flows, etc. Erosion *per se* is therefore inevitable. What is worrisome is the relationship between the rate of erosion and that of soil formation which humans affect, often negatively.

Sustainable development

Sustainable development has been defined by the World Commision on Environment and Development (1987)

> as a process of change in which the exploitation of resources, the direction of investments, the orientation of technological development, and institutional development are made consistent with future as well as present needs.

Amongst other things, the concept requires the preservation and enhancement of the productive capacity of natural resources; an equitable land distribution; a redistribution of wealth to the poor; and the maintenance of natural ecological processes.

Any human intervention in nature will only change the dynamic interactions that predominate. Consequently, the only land-use policy that will produce zero additional pollution, erosion or species extinctions is one that keeps humans off that land. Since we obtain our sustenance from the land, this is clearly impossible, other than perhaps in some specially designated areas.

Techniques may be devised to reduce erosion and species extinction, such as levelling of land, construction of terraces, creating of hedges and natural refuges, and special methods of cultivation. However, these methods are energy consuming and, even though they reduce erosion, they produce pollution. Georgescu-Roegen (1971) maintains as a law of economics that every human activity by necessity creates pollution in the same manner that work increases entropy. We agree with such a position. Devising a sustainable land-use policy that does not alter the land is therefore impossible; every use degrades the system compared to its original condition.

However, it may be feasible to create such policies if we are willing to accept a new equilibrium between pollution, erosion and production.

Combating land degradation

Humans use land primarily to obtain food and fibre, to obtain materials with which to build shelter and other objects, and for recreation. The intensity of these activities is the main determinant of the rate of land degradation. So, for example, hunter-gatherers degrade land much less than agriculturalists; traditional villages affect an area less than modern cities; and so on. The rate of degradation is also mediated by technology. Land can be degraded less under the same intensity of use if appropriate techniques are used to minimise negative effects. So, for example, drip irrigation technology may cause less damage than flood irrigation and conserve more water. More intense land use may also provide more benefits: yields may be higher and more people fed by modern agriculture than by slash and burn methods. Similarly, more people can be housed and more services provided per unit of land in a modern city than in a traditional village. Devising a land-use policy is therefore a question of determining the best way to satisfy human needs with the least environmental (erosion, species loss, pollution) and energy costs.

For the purposes of this analysis, we can classify land degradation into two kinds which we will call "direct" and "indirect". The first is the type of degradation that is immediately visible and affects the quality of the production on the land. If plant cover is removed soil erosion increases; if cattle or other grazers are introduced, the abundance and biomass of palatable grass species decreases; if a forest is clear-cut, some species may become locally extinct; if part of the production of an area is exported, be it in the form of a plant crop, animals, or forest products, soil nutrient stocks will diminish. Because the negative impacts are very apparent, societies try to prevent or minimise them. Scientific agriculture tries to devise techniques that will lessen the negative impacts of farming. Nevertheless, most of the best-known cases of land degradation involve these direct effects: the desertification of Cartago; salinisation of the Tigris and Euphrates river basins; the "dust bowl" areas of the central United States; desertification in the Sahel, etc.

The indirect type of negative effect is one of which we have been much less conscious until recently. Increases in flooding down-river as a result of the removal of vegetation up-river; changes in evapotranspiration with changes in vegetation; global increases in air and water pollution resulting from developments in the chemical industry; the planetary extinction of species, all provide examples. Since these costs are not borne – at least initially – by the users of the land, they have tended to be ignored. And because rivers, oceans and the atmosphere have some degree of buffering, it has been assumed that they can be used as dumping grounds without much problem. When the numbers of humans were low, such behaviour was acceptable. But with the increase in human density, and the virtual occupation of every area of the world, there is always somebody who is affected negatively by land degradation.

Reducing both direct and indirect degradation costs must be the aim of any rational land-use policy. Nobody consciously tries to degrade land, but it is an inevitable consequence of use. Social scientists (Foweraker 1981) have determined that when humans occupy a new area (a so-called agricultural frontier) their attitude towards land use goes through three stages. In the first stage, the land, which at that point in time is plentiful, is treated as if nothing one does to it matters. Consequently, if the area is a forest, trees are cut down indiscriminately and only a few are used; cattle are introduced to savannas but not tended adequately until there are too many; and if the area is good for agriculture only, the minimum effort is made to increase productivity. Land at this stage is seen as a free good, and therefore not worth investing money or effort to keep it from degrading. As time goes by, and as all available land is occupied, it ceases to be a free good, and a phase of legal occupation and ownership takes place. The value of land increases at this second stage. Land-use practices improve, but techniques are still primitive. The last stage is the technification of land use, when methods that almost invariably involve more energy inputs, and consequently are more costly, are introduced. Therefore they can only be justified if the corresponding yields also go up.

We assume that owners of land wish to increase the profit or rent that they derive from the land. It is not in their interest to degrade the land and reduce their income. However, they may not know any other way of using their land; or they may lack the capital needed to increase their yields; or they may have determined that the increases in yields are less than the increases in costs, either because costs are very high or the value of the land is very low, and therefore it is not economical for them to invest the needed capital; or even when it is economical there may be other investments that are more profitable or less risky, and therefore it is in their interest to use their money that way. Another reason for not investing in land improvements is an unpredictable political climate, which makes any investment risky. All of these situations are present now, or have been in the past, in one or another savanna country.

There is very little unused land left in the world. Also the human population is growing, and so are aspirations in terms of living standards. It is therefore important to devise strategies that will increase yields and efficiency of land use with the minimum possible degree of direct or indirect degradation. The best strategies will vary from area to area, and will no doubt themselves be subject to evolution as time goes by. They will be closely tied to scientific knowledge that is constantly increasing; to the amount of capital available; and to the value of land, and of the products that are taken from that land. This applies especially to savannas.

INCREASING PRODUCTION OF TROPICAL SAVANNAS

Savannas are among the areas of the world with the greatest physical potential for increasing agricultural and animal production. Yet, until recently and excepting some parts of Africa and India, savannas have been used extensively for cattle-raising or not at all. African savannas, on the other hand, have been used by pastoralists and shifting agriculturalists for a very long time. Indian savannas are largely secondary savannas, the result of human intervention in an originally forest vegetation. This situation is now changing very fast and savannas, especially in Africa, are increasingly being occupied with the objective of using them for agriculture. In order to plan ways to increase yields with the lowest possible environmental costs, it is necessary to learn which ecological and economical constraints have a bearing on savanna regions. The type of information needed is as follows:

(1) Some historical context. When was the area first occupied; what has its land-use history been; how much change has taken place in the environment, and in the human society? History of land ownership patterns.

(2) Present land use. What agricultural and grazing techniques are used; what effect are they having on the soil, the vegetation, the fauna? How profitable are they? Who benefits most from present land use.

(3) Effect of national and international policies on land use. Government subsidies; credit availability; tariffs on agricultural products; internal and export markets; choice of products; exchange rates.

(4) Agricultural techniques. What techniques are available; how educated is the rural population; do they have access to agronomical advice; do they have access to land, to capital; what are the labour costs and availability? What ecological effects do these techniques have; are there more efficient alternatives available, and why are they not used? Where are the markets?

(5) Other land uses. What is the potential to develop the landscape for tourism; or as areas that attract money for conservation?

16

Unfortunately, when all this information is drawn together into a coherent framework, the real socio-economic potential of many savannas is much less than a cursory evaluation would suggest.

The issue and case studies

In the following pages a series of studies relating to particular issues and regions is presented. Ideally we would have liked to have had all the above information available for all the savannas chosen as case studies. This has not always been possible. Instead, authors have concentrated in each instance on a particular feature that they feel is the principal independent variable. For example, the effect on savanna development of the price of oil in Venezuela (Dutch disease), or the consequence of government subsidies in Brazil or of internationally set, artificial prices of beef and their effect on land use in Africa. What emerges is a confirmation of the diversity of the physical and biological conditions of savannas and of their uses. Nevertheless, as identified in the last chapter, many common themes and general recommendations do emerge.

REFERENCES

Adámoli, J. and Fernandez, P. (1980) Expansión de la frontera agrícola en la Cuenca del Plata: Antecedente ecológicos y socioeconómicos para su planificación. In: O. Sunkel and N. Gligo (eds) *Estilos de desarrollo y Medio Ambiente en la America Latina.* Fondo de Cultura Económica, México, pp.468–501.
Bourlière, F. and Hadley, M. (1983) In: F. Bourlière (ed) *Tropical savannas.* Ecosystems of the World vol. 13. Elsevier, Amsterdam.
Brown, L.R. (1981) *Building a sustainable society.* W.W. Norton, New York.
Fernside, P.M. (1986) *Human carrying capacity of the Brazilian rainforest.* Columbia University Press, New York.
Foweraker, J. (1981) *The struggle for land.* Cambridge University Press, Cambridge.
Georgescu-Roegen, N. (1971) *The entropy law and the economic process.* Harvard University Press, Cambridge.
Hawley, A.H. (1986) *Human ecology.* University of Chicago Press, Chicago.
Meggars, B.K. (1954) Environmental limitations on the development of culture. *American Anthropologist* 56:801–24.
Meggars, B.K. (1957) Environment and culture in the Amazon Basin: An appraisal of the theory of environmental determinism. Anthropological Society of Washington, Studies in Human Ecology, Pan American Union, Washington, DC, pp. 71–90
Norgaard, R.B. (1987) Economics as mechanism and the demise of biological diversity. *Ecological Modelling* 38:107–122.
Oelschlaeger, M. (1991). *The idea of wilderness.* Yale University Press, New Haven.
O'Neil, R.V., DeAngelis, D.L., Waide, J.B. and Allen, T.F.H. (1986) *A hierarchical concept of ecosystems.* Princeton University Press, Princeton.
Ratzel, F. (1882–1892) *Anthropogeographie.* J.Engelhorns, Stuttgart.
Semple, E.C. (1911) *Influence of the geographical environment.* Holt, New York.

Sober, E. (1986) Philosophical problems for environmentalism. In: B.G. Norton (ed) *The preservation of species.* Princeton University Press, Princeton. pp. 173–94.

Solbrig, O.T. (1991) Ecosystems and global environmental change. In: R. Correll (ed) *Global environmental change* Springer-Verlag, Berlin, pp. 97–108.

World Commission on Environment and Development (1987) *Our common future.* Oxford University Press, Oxford.

Section 2

Policy principles

CHAPTER 2

ECOLOGICAL CONSTRAINTS TO SAVANNA LAND USE

O.T. Solbrig

SUMMARY

Two concepts have dominated the thinking regarding ecology and land use: (1) that ecosystems are equilibrium systems and (2) that people are an outside source of disturbance. Savanna systems, however, are more appropriately viewed as non-equilibrium systems within which people form an integral part. When disturbed, such systems are likely to adopt an entirely new stable or quasi-stable state. Another characteristic of non-equilibrium systems is that they may oscillate continually due to their own internal dynamics. These oscillations can be regular (limit cycles) or irregular (chaotic oscillations). These characteristics of non-equilibrium systems have important management implications. In effect, non-equilibrium degraded ecosystems do not necessarily return to their previous state when the source of disturbance is removed, and in most cases it might not be possible to return them to their original condition. Furthermore, it also means that one management regime will not suffice for all savannas. The previous history and present disturbance regime will determine the outcome of management systems. Finally the levels of human investment – in capital, labour and knowledge – in land use are major determinants of the behaviour of managed savanna systems.

The lesson is that no single set of general management prescriptions can be offered for the "proper" use of savannas. The level of human investment, the objectives of the human society, and the characteristics, as well as the previous history, of a savanna will determine which management regimes satisfy human aspirations and which do not.

INTRODUCTION

In this chapter some of the ecological characteristics of tropical savannas and some of the human activities that take place there are reviewed. It describes the physical characteristics of savannas, their biological attributes, the traits of the human societies living there, and their land-use practices, placing special emphasis on the technologies employed. By necessity the analysis is not exhaustive, but will concentrate on the principal types of savannas. More details are provided in the case studies that follow.

First, the principal determinants that condition a tropical savanna ecosystem are described and some of the ecological controls on economic development presented. This is followed by a brief discussion of the meaning of ecological equilibrium. Several types of savanna land use are then described, pointing out that savanna structure and function and human societies are modified in each case. It concludes with some general thoughts about conservation and development.

DEFINITION AND CLASSIFICATION OF TROPICAL SAVANNAS

Tropical savannas are very heterogeneous ecosystems, and consequently are hard to define. Savannas consist of a series of ecosystem types which form a distinct biome (Huntley and Walker 1982; Sarmiento 1984). Following Sarmiento (1984; Frost et al. 1985) a savanna is defined as

> an ecosystem of the warm (lowland) tropics dominated by a herbaceous cover consisting mostly of bunch grasses and sedges that are more than 30 cm in height at the time of maximum activity, and show a clear seasonality in their development, with a period of low activity related to water stress. The savanna may include woody species (shrubs, trees, palm trees), but they never form a continuous cover that parallels the grassy one.

This definition encompasses a great diversity of landscapes, from dry grassy scrublands with low average (300–500 mm) and irregular rainfall such as in the Sudano-Sahelian region of Africa, to open woodlands with more than 1500 mm of yearly precipitation as in many parts of South America. It encompasses ecosystems with relatively fertile soils as in the Serengeti grasslands of Tanzania and Kenya, and some with extremely infertile soils as in parts of northern Australia. The term is applied also to ecosystems with a temperature seasonality, such as those in Zimbabwe and South Africa and the Chaco region of South America. This broad interpretation of what constitutes a tropical savanna confuses ecological generalisations and confounds the interpretation of land-use practices.

22

According to Sarmiento (1983) tropical savannas can be divided into four distinct functional types:

(1) *Semi-seasonal savannas*. These are savannas with weak water stress conditions or responses, with low levels of plant dormancy or adaptations to fire (called semi-evergreen seasonal forests by Cochrane et al. 1985).

(2) *Seasonal savannas*. Characterised by a distinct rainfall seasonality, they are moulded by an alternation from a wet to a dry season that is reflected in plant growth. During the dry period, fire is a normal and regular event. This is the most widespread type of savanna and includes most of the *cerrado* of Brazil, the *llanos* of Venezuela and Colombia, the eastern and western savannas of Africa, and the savannas of northern Australia. (In this book we deal exclusively with issues that relate to seasonal savannas. These are in any case the most widespread kind of savanna.)

(3) *Hyperseasonal savannas*. These are savannas where, in addition to a well-marked dry season and accompanying water stress, there is also a period of flooding during the wet season that creates a different type of stress (also known as 'poorly drained savannas'.) An example of a hyperseasonal savanna is the Brazilian *pantanal*.

(4) *Esteros*. These are areas without a clear dry season, but with an excess of soil water for most of the year. These are not generally considered to be typical savannas.

Geographical distribution of seasonal savannas

Seasonal savannas are well represented in Africa, where savanna and savanna woodlands cover more than half of the total area of the continent (see Figure 1.1). African savannas are extremely diverse. Rainfall varies from 300 mm/year to over 1500 mm/year. Soil quality varies from very poor to fairly good. This environmental diversity is reflected in the variety of vegetation types, from dry grasslands in areas of low precipitation all the way to tall woodlands in wet savannas, and land uses from nomadic pastoralism to agriculture and tourism.

Much of the Indian subcontinent is covered by a rather uniform savanna vegetation, believed to be largely the result of human disturbance. Areas of Thailand, Cambodia and Vietnam are covered by open deciduous dipterocarp forest, which is also considered a type of savanna (see Figure 1.1; Stott 1984).

In Latin America savannas occupy much of the Orinoco basin (the so-called *llanos* of Colombia and Venezuela) and the high central plateau of Brazil (*cerrados*). Smaller pockets of savanna are found in Bolivia (*llanos de Moxos*) and in the Paraguayan and Argentinian Chaco (Sarmiento 1983, 1984). In the Amazon basin savanna vegetation is found in islands of extremely poor sandy soil. Pockets of savannas are also found in the Guyanas. Rainfall in the South American savannas is higher than in many African savannas, with precipitation usually in excess of 1000 mm. Soil texture is poor and nutrient levels low.

Savannas occupy the dry tropical areas of northern Australia (Bourlière 1983). They are intermediate in moisture between African and South American savannas and generally have very poor soils (Holmes and Mott, this book). Land use in savannas varies from hunting and gathering, through extensive cattle-raising, to agriculture in the wetter areas. Tourism, recreation and conservation also occupy a significant proportion of the landscape. Much of the infrastructure has been financed by mineral development.

The following discussion will refer only to the seasonal and semi-seasonal savannas which are the most abundant and distinctive types of savannas in the tropics.

DETERMINANTS OF SAVANNA STRUCTURE AND FUNCTION

Productive processes of savannas (using productive in its biological sense) are affected primarily by an oscillating and variable rainfall pattern, by low soil nutrients, by herbivory, and by regular burning. Light and temperature are uniform and predictable in most savanna areas.

Water

The grasses and sedges (and other herbaceous plants) in savanna ecosystems control and regulate ecological processes such as water balance, productivity, mineral cycling, fires and herbivory. Although this stratum is referred to as continuous, it only is so at the height of its growth, since the area covered by the bases of the grasses may only be 10–20 per cent of the total. Although the common grass species of savannas have wide geographical distributions, each species has a distinct phenology and microdistribution.

Water is the key environmental variable in savannas. Soil moisture regimes are affected by:

(1)   The quantity and seasonal distribution of annual rainfall and the proportion of this water that enters the soil;

24

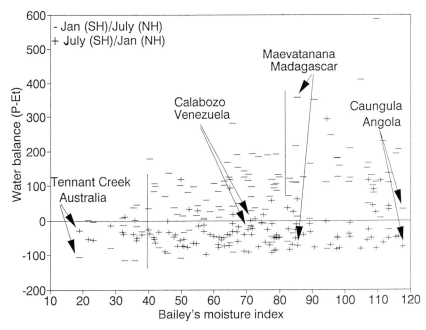

Figure 2.1 Relation between Bailey's moisture index (that corrects total precipitation for temperature effects) and seasonal water balance (precipitation minus evapotranspiration) for a sample of over 200 savanna sites. Note (1) the great diversity in the moisture index and (2) the positive water balance in savannas during the wet season, and negative during the dry season

(2) By the water-holding capacity of the soil, which is a function of soil texture and depth; and

(3) By the amount of evapotranspiration, which is related in complex ways to climate, soil texture, soil surface characteristics and the type of vegetation at a site (Frost et al. 1985).

Mean annual rainfall in tropical savannas extends from about 300 mm to over 2000 mm and normally does not exceed yearly potential evapotranspiration, though it does so seasonally (Figure 2.1). The vegetation of tropical savannas develops in two contrasting rhythms during each annual growth cycle: a period of intense and diversified growth during the rainy season and a cycle of more or less prolonged low activity. This growth seasonality is reflected in the changes in form and function of the ecosystem throughout the year (Silva 1987). Except in areas lying near or within the subtropics, that is, where temperature has a pronounced seasonality, there is no true dormancy among savanna grass species (Sarmiento 1984).

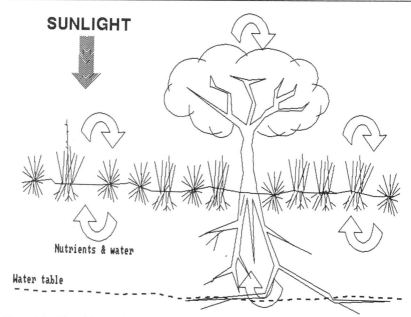

SUNLIGHT

Nutrients & water

Water table

Figure 2.2 Pictorial model of savanna function. Solar radiation is plentiful for both grasses and trees. Grasses circulate nutrients and water from the upper layer of the soil. Because of the negative water balance during the dry season, grasses must then reduce their activity considerably. Trees with their deep roots have access to water in the water table during the dry season. However, lower soil layers are poorer in nutrients, requiring a more extended root system

A distinctive feature of savannas is the coexistence of herbaceous and woody plants. Characteristically the woody species grow dispersed amongst a more or less continuous layer of herbs, mostly C4 grasses (C4 refers to a type of very efficient photosynthesis distinctive of many tropical plants). This coexistence has been hypothesised (Walter 1969; Walker 1985; Frost et al. 1985) to be the result of a partition of the soil substrate into two horizons: a superficial one dominated by the roots of herbaceous plants, and a deep one controlled by the roots of trees (Figure 2.2). According to this theory, the importance of trees diminishes when the water table is very deep, or when soils become shallow, as in the case of a well-developed lateritic hard pan (Santamaria 1965; Santamaria and Bonazzi 1963, 1964) that restricts the growth of deep tree roots. The first case is common in high plateaus; the last is typical of alluvial plains. Thus, in areas with low precipitation, or very good drainage, trees are absent or restricted to depressions where water accumulates (Sarmiento and Monasterio 1971), while in wet areas with poor drainage trees are found only on elevations. That is, the dynamics of water in the landscape controls the structure of the community at this first physiognomic level. The structure and composition of plant and animal communities along moisture gradients appear highly correlated with soil-water dynamics (Silva 1972; Silva and Sarmiento 1976a,b).

Soil characteristics

The soil is the substrate on which plants grow and from which they obtain water and nutrients. The physical characteristics and the underlying geological formation determine both the nutrient content and the water and nutrient retention capacity of a soil.

Savanna soils vary widely in texture, structure, profile and depth, reflecting the geology, geomorphology and climate, and the influence of topography, relic features of past landforms, the kind and extension of vegetation cover, and animal activity of the particular savannas (Young 1976; Montgomery and Askew 1983; Frost et al. 1985). Three other factors also play an important differentiating role in soil formation: topography, parent material, and age.

The American, Australian and West African savannas are found on soils that are extremely poor in nutrients and which reflect the high temperature and moisture regimes of these regions. Most savannas occupy plains of relatively flat or slightly undulating topography. The principal influence that relief has over the ecosystem is on regulation of the drainage conditions and, ultimately, on the water balance. This influence in turn translates into important consequences regarding the chemical and nutritional characteristics of the soils. The factors that produce the relief, through their action on soil formation, indirectly determine the physico-chemical characteristics of the soils (Sarmiento 1984).

Soil nutrients

Savanna soils generally have a low reserve of weatherable minerals. The effective cation exchange capacity of savanna soils (a measure of the capacity of a soil to provide nutrients to plants) is generally low. They have small amounts of exchangeable calcium and magnesium (Jones and Wild 1975; Lopes and Cox 1977; Mott et al. 1985; Frost et al. 1985). Phosphorus levels are sometimes extremely low, and some soils rich in sesquioxides (such as in the American savannas) have a high capacity for fixing phosphorus (which consequently becomes unavailable to plants). Some highly weathered soils also have high levels of exchangeable aluminium (Lopes and Cox 1977) which can be very toxic, especially to crops and introduced forage grasses (Medina 1978; Sarmiento 1984).

27

The process of mineralisation and the circulation of nutrients in the soil are incompletely understood. A better grasp of this process is crucial for a more rational use of savanna soils. The nutrient status of soils is related principally to the age of sediment deposition. The poorest soils (oxisols and ultisols) are those derived from the oldest deposits, since these materials have been subjected to soil-forming processes for prolonged periods of time. There exists a large number of studies that contain information on soil nutrient status of American and Old World savannas (Zinck and Urriola 1970; Guerrero 1971; Guerrero and Cortes 1976; Leon and Botero 1980; Torres 1980; Montgomery and Askew 1983; Sarmiento 1984; Goedert 1986; Medina 1987). These studies agree in classifying the soils of most American, Australian and some African savannas as oligotrophic, acid, and very poor not only in macronutrients but also in many other essential nutrients, such as sulphur (Avilan and Rojas 1975; Gualdron and Salinas 1982). Furthermore, the American savanna soils are high in aluminium (Vargas 1964), considered toxic especially for crops.

With the exception of savannas with extremely acid soils, the amount of organic matter is the main determinant of the cation exchange capacity of soils. In wet savannas, high rainfall and an extended wet season favour increased plant production with a consequent input of organic matter into the soil. As fire effectively mineralises most of the aerial matter produced (Sanford 1982; Menaut et al. 1985), organic matter input is almost exclusively the result of increased below-ground production (mostly roots and root exudates). However, microbial activity is limited by the low levels of assimilable carbon, high C:N ratios, high lignin content in the vegetation, and in some cases high amounts of condensed tannins and secondary chemicals.

Soil erosion

The severity of soil erosion is mainly determined by climatic and soil factors and topography (Hudson 1975). The ease with which soil particles become detached depends on soil structure and texture. Erosion can take place both during the dry and the wet season. During the long dry season the soil is liable to blow away because of the incomplete cover of vegetation. Erosion is worse where cattle, goats or sheep graze the vegetative cover; where the woody species are browsed heavily by goats or camels; where people remove trees for firewood; or where the vegetation has been burned early in the dry season. In humid and sub-humid savannas, most erosion occurs when heavy showers and thunderstorms fall on soil with an incomplete cover of vegetation (UNESCO 1979).

In grazed savannas erosion can be reduced by controlling grazing animals

Table 2.1 Yields of cowpea and intercropped cowpea and maize under different tillage methods. All plots received 30 kg of nitrogen per ha 30 kg phosphorus per ha and 30 kg of potassium per ha (from UNESCO 1979)

	Tillage method			
Grain yield	Ploughed, ridged	Ploughed, flat bed	Strip	Zero
Sole cropping				
Cowpea	1185	1274	1538	1649
Intercropping				
Cowpea	665	725	1022	941
Maize	1705	1675	2337	2809
Cowpea & maize	2370	2400	3359	3750

and restricting burning to the end of the dry season, so that as much plant cover as possible is retained at all times. In cultivated savannas, the use of mixed and relay cropping techniques can keep plant cover over the soil for most or all of the year, and is a good way to reduce erosion. Yields in excess of those from planting only one crop can be obtained (see Table 2.1), and such methods may be particularly advantageous when combined with zero tillage techniques. The full potential of mixed and relay cropping has yet to be realised (UNESCO 1979).

Fire

Regular fires are one of the universal events in tropical savanna ecosystems. Although lightning was and occasionally still is the source of spontaneous fires (Komarek 1972), today most fires are human made. Fire has a profound effect on savanna structure and growth, to the point that some researchers feel that the grass/tree physiognomy of all tropical savannas is the result of fire. Most students of savannas reject this extreme view. Yet it is clear that in many regions, such as India, savannas are the result of human action, especially burning.

Coutinho (1982), Gillon (1983) and Frost and Robertson (1987) have reviewed the ecological effects of fire on savannas. Fires reduce standing biomass and litter, and kill individual organisms, seeds, seedlings and unprotected plant tissues. Fire also changes the energy, nutrient and water fluxes between soil and atmosphere. The nutrient status and productivity of savannas are determined in part by the frequency, timing and intensity of burning.

Herbivory and disease

Ticks are the major parasites of livestock in savanna regions all over the world. They cause blood loss, severe discomfort and injection of toxins. Other animal pests and parasites of both cattle and humans (principally malaria), as well as tropical weeds, are also important in limiting the use of tropical savannas. Chagas' disease (due to *Trypanosoma kurtzii*), transmitted by the *vinchuca* or *chipo* (*Triatoma* sp.), limits human work capacity in the American savannas. In many African savannas sleeping sickness (trypanosomiasis), which is transmitted by the tsetse fly (*Glossina* spp.), and rinderpest virus disease constitute an additional serious limiting factor in stock-raising.

ARE SAVANNA ECOSYSTEMS IN ECOLOGICAL EQUILIBRIUM?

Usually it has been assumed that natural ecosystems are in thermodynamic equilibrium. By this is meant that, if not interfered with by humans, they will persist in their present state for long periods of time (DeAngelis and Waterhouse 1987). Since natural ecosystems are affected by all manner of natural disturbances (droughts, floods, fires, storms, etc.), equilibrium hypotheses predict that the system will return to its previous state once the source of the disturbance has disappeared. It also implies that they will return to their present state after a human-made disturbance.

Technically equilibrium hypotheses imply that there are negative feedbacks in the ecosystem that control the behaviour of the various elements. So, for example, plant growth is supposedly controlled by herbivore pressure, and by decomposition rates that control nutrient availability in the soil; herbivores in turn are supposedly controlled by the availability and quality of plant biomass, and/or by predation rates, and so on.

Theories about the optimal way to manage ecosystems have been dominated by this way of thinking. So, for example, the introduction of domesticated herbivores is seen as a force that moves the system away from equilibrium, which will inevitably result in a reduction in the availability and quality of the plant resources required by the grazing animals. To compensate for this situation an effort is made to eliminate wild herbivores (which are seen as competitors) and to "improve" the quantity and quality of the vegetation through artificial means, such as fertilisation, control of stocking rates, periods of rest for the range, irrigation, etc.

Increasingly a different sort of thinking is influencing the way ecosystems are perceived. This new thinking is represented by the so-called "non-equilibrium" hypotheses (Holling 1973; Noy-Meyer 1975; Wiens 1977, 1984;

DeAngelis and Waterhouse 1987; Westoby et al. 1989). These hypotheses maintain that natural ecosystems are not in equilibrium but instead are the result of various historical and chance factors, and that a system such as a savanna can be present in multiple states. Studies with non-equilibrium systems (Nicolis 1991) indicate that their behaviour is very different from that of equilibrium systems. Most notable among the characteristics of their dynamics is the existence of "bifurcations". That is, at particular points in their evolution, non-equilibrium systems can assume more than one equally probable and equally stable state (Nicolis and Prigogine 1977, 1989). Non-equilibrium systems are therefore inherently non-predictable. The same disturbance can create one state in one instance and a very different one in another case. The very mixed results obtained in different savannas when oversown using identical techniques may be an example of non-equilibrium. If this is true then the existence of a savanna in any one of a multitude of states may indicate that it is being managed sustainably. This considerably complicates management. It requires a much more careful analysis than what has been undertaken in most instances to determine which model is best suited for a particular system. One important consideration is the relative importance of internal biotic factors such as negative feedbacks between plants and herbivores, and external factors such as droughts and storms.

There is also a yearly variation in rainfall which imposes an additional stress. The variance in interannual precipitation is inversely related to total rainfall (Bailey 1966; Bailey et al. 1977). Furthermore, the lower the average rainfall, the more extreme the effect of drought years on the flora and fauna. Consequently, dry savannas are almost certainly non-equilibrium systems (Ellis and Swift 1988), while it is less certain of wet savannas.

What are the ecological and management implications of non-equilibrium systems? According to Westoby et al. (1989) savanna dynamics can be described by a set of alternative stages of the vegetation and a set of discrete transitions between states. Consequently, they recommend that the manager make an inventory of the possible alternative states of the system and of the possible transitions from one state to another, remembering that there is more than one possible transition. Observant managers (such as nomadic pastoralists) usually possess such information. This knowledge will allow the manager to develop adequate responses for each state and possible transition.

SAVANNA LAND USE

Savannas are among the oldest ecosystems used by people. The earliest remains of humans have been found in central Africa in areas now covered with savannas. Anthropologists believe that these human populations existed by hunting wildlife and gathering plant materials. Some traditional cultures of savanna hunter-gatherers (e.g. Australian Aborigines) still survive. Present-day savanna land use by both traditional and non-traditional cultures primarily includes grazing livestock, harvesting wood, and growing crops. Because hunting and gathering as a form of savanna land use is of little global importance, it will not be considered further.

The management of savannas is concerned primarily with manipulating the two principal limiting factors to which seasonal savannas are subjected: water and nutrients. Water shortages are seasonal: a long season without available soil water for the herbaceous vegetation with more or less shallow roots. This environmental limitation and the corresponding drought stress it induces on savanna grasses are very hard to overcome. Where there are possibilities for irrigation, it cannot be justified economically to maintain a natural grassland, since irrigation offers more lucrative alternatives: improved grasslands or crops. Nutrient limitations are severe. To correct them implies reducing soil acidity by liming followed by applications of phosphorus and nitrogen fertiliser. The costs of such treatments are high and cannot be justified in order to maintain livestock productivity in natural grasslands.

Seasonal savannas can be managed at many contrasting levels of technology and production. One system is that of the traditional extensive livestock grazing or traditional agriculture which does not employ many inputs (fertilisers, pesticides, energy, machinery, improved varieties, etc.). This is the most elementary way of utilising the savanna, the least costly, and the least productive. It is also highly sustainable, having been in operation in Africa for at least 6000 years. The other extreme is represented by a system of high inputs that includes the replacement of savanna vegetation by introduced grasses or by crops. Without irrigation only one annual crop (corn, sorghum, cotton, etc.) can be obtained in the seasonal savanna, and only through the use of species that are resistant to drought (such as *Panicum maximum* or *Hyparrhenia rufa*) and not too demanding regarding nutrients can a permanent artificial grassland be maintained. In both these last cases the investment is relatively high, as is the resultant primary and secondary productivity. The sustainability and economic viability of high-input systems has been questioned.

Between these two extremes we have a large diversity of traditional and non-traditional systems of medium-intensity management, which involve simple technologies (burning, differential grazing) low levels of inputs (fertilisation, some energy), that can lead to a notable increase in productivity in comparison with the extensive system. Such systems may be more sustainable than high-input systems. Annual rainfall is a major factor in determining whether agricultural systems are viable in savannas. As a rule of thumb, savannas with less than 600 mm of average yearly rainfall can not sustain a regular cropping regime in either an economic or a physical sense.

We now describe in very general terms various savanna management systems, giving specific examples and emphasising the ecological and economic constraints of each of these approaches. Savanna land use can be classified in many different ways according to what criteria are used. For the purposes of this presentation we have divided savanna land-use systems into three groups. These should not be interpreted as a historical sequence, nor as representing levels of optimality. There are circumstances where each of these types of land-use is more appropriate than the others. As indicated in the case studies that follow this chapter, the physical and biological characteristics of the area and the social structures of the societies using the land will determine which system is the most suitable for each region.

EXTENSIVE GRAZING AND SIMPLE AGRICULTURE

Historical overview

Homo sapiens evolved in a savanna environment in East Africa about one million years ago. For most of its history our species has been associated principally with tropical savanna environments, obtaining its subsistence from hunting game and gathering seeds, fruits and roots. Herding and agriculture are a relatively recent development (about 10,000 years ago) and occurred first in areas outside tropical savannas.

Extensive herding of ruminants, which today represents the most widespread way of using dry seasonal savannas, was first practised in Africa about 6000 years ago. Herding probably diffused from the Middle East where it began about 10,000 years BP. African nomadic pastoralism developed in an almost continuous band from the Atlantic to the Indian Ocean (Fulani, Galla, Nilo-Hamites). African pastoralism is based on the use of cattle, camels, sheep and goats, with burden animals playing only a minor role. The productivity of this type of extensive stock-raising is often very low but has proven to be ecologically sound. In areas where this traditional land use has been practised for a long time efficient techniques have evolved that increase yields and protect the soil from erosion (Little and Leslie 1990; Odegi-Awuondo 1990; see also Lane and Scoones, and Mentis and Seijas, this book).

Agriculture is possible in tropical savannas whenever rainfall exceeds 600 mm and the rainy season is more than 3 months long. A great diversity of agricultural and agro-pastoral societies and cultures developed in different Old World tropical savannas. Agriculture diffused into the savanna regions of the Old World from further north (Middle East, Ganges valley, Thailand, north-central China), but Africa (particularly the Ethiopian highlands) became a centre of plant domestication (Hawkes 1983).

Agriculture is often linked with pastoralism in Africa, and traditional societies there have acquired a great deal of knowledge and experience in this type of land use (Jones and Wild 1975; Lane and Scoones, this book). Three stages of agriculture can be distinguished. Where population is low, a very extensive type of shifting cultivation is practised, often with movement of village sites. As population density increases, rotation is used much more systematically. Where population density is very high, continuous cropping takes place. Without some use of fertiliser, soil degradation accompanies the intensification of agriculture.

An alternation of fallow and cultivation is a common practice in traditional African savanna agriculture (Pelissier 1960; Nye and Greenland 1960). Another common practice is mixed and sequential cropping, which reduces the risk of complete crop failure. This type of cropping spreads the demand for nutrients over a longer period, minimising the peak demand; it also reduces the erosion hazard.

In pre-Columbian South America, the savanna was occupied primarily by hunter-gatherer societies, although there were some agricultural societies. But the Spanish and Portuguese colonialists used the savanna purely for pastoral activities, restricting agriculture to wetter forest and coastal areas. Livestock spread very rapidly in Latin America after its introduction by the Spaniards early in the sixteenth century. Cattle reproduced so fast that native communities at first saw them as a great nuisance (Chevalier 1975). Both the Brazilian *cerrados* and the Venezuelan *llanos* emerged as the two great cattle regions. In seventeenth-century Brazil, cattle so dominated life along the Sao Francisco River that Joao Capistrano de Abreu described it as the "era of leather" (Abreu 1954). The native populations developed a taste for beef. It is reported that in the eighteenth century Caracas consumed half as much beef as Paris, even though it had only one-tenth the population (Azara 1954).

In Australia, the story is similar but of more recent occurrence. Until the middle of the nineteenth century, Australia's savannas were inhabited by hunter-gatherers who used fire as a means of manipulating wildlife populations. English colonialists, however, were keen to establish livestock grazing and, whilst permitting the Aboriginals to live on the land, soon made cattle-raising the dominant activity. Today hunting plays only a minor role in Australia (but see chapter by Holmes and Mott).

Animal domestication encouraged the intensification of land use and a higher intensity of occupation. This specialisation took place first in Africa on the basis of a division of labour between generally sedentary farming societies and nomadic herding societies.

Pastoralism

African pastoralism is characterised by the collective ownership of land (see chapters by Lane and Scoones, Kituyi, and Mentis and Seijas, for more details) and individual pastoralist family ownership of livestock. The livestock is kept for milk, meat and hides. A family normally owns an average of 50–100 head of cattle or the equivalent. It has been estimated that a family of six to eight persons needs only 15–20 cows or the equivalent to survive. The larger herd is insurance for years of drought, against disease epidemics, or to withstand raids on the herd. Large herds are also needed for bridewealth and as a source of prestige. Most herds are mixed, but cattle are held in much greater esteem than sheep, goats or camels.

Nomadic pastoralism is a management system which copes with the alternation of excess and shortage of grass and water created by the alternation of dry and wet seasons. Livestock is herded to areas where there is fodder. During the rainy season when there is usually ample grass, the herd is allowed to roam in areas of the savanna that have little grass during the dry season, obtaining its water from rain puddles, while during the dry season it is kept in the vicinity of permanent water holes. By not grazing an area continuously, pastoralists allow the range to recover. The herd is often divided into a home herd consisting of lactating cows, some bulls and young animals, tended by old men, women and children, and a *fora* herd of bulls, non-lactating cows and young animals which is driven by the young men to remoter areas with better grass. The precise composition of the herds and the systems of grazing vary from group to group and represent adaptations to different ecological conditions.

African pastoralists do not have specific strategies to preserve the floristic composition of the range nor do they avoid overgrazing. Yet their movements are so timed that, when the range becomes overgrazed, they migrate to a new area before permanent degradation takes place. The major impetus for these movements comes from seasonal changes. In this they follow strict rules enforced by tradition. This system of rotation allows the savanna to recover. Similar techniques (intensive grazing for short periods followed by a long period of recovery) are advocated in modern ranching, but are not widely practised because of the increased costs of fencing and labour they entail (see chapter by Holmes and Mott).

Unfortunately, with the inclusion of pastoralist societies within the borders of national states, their freedom to migrate has been curtailed severely. So, for example, the Turkana of northern Kenya can no longer move their *fora* herds to the mountains of Ethiopia as they have done for centuries, if not millennia. Another example is the use of part of their territory for agriculture, as in the case of the Barabaig of Tanzania (as explained in the chapter by Lane and Scoones). Restricting the herds to a smaller territory forces pastoralists to stay in an area for longer periods which has resulted in overgrazing and range degradation. Furthermore, there has been a tendency for pastoralists to become permanently settled in an area, a change that is encouraged by national governments. This further restricts the movement of the herds and increases overgrazing and range degradation.

Modern veterinary medicine increases herd sanitation and survival. Artificial water wells increase survival of the herds during dry years. By eliminating the natural controls on herd size, pastoralists today maintain larger herds than they would traditionally and do so within a smaller territory. The larger herds and the more sedentary nature of the pastoralist population result in overgrazing and soil erosion. In turn the poor condition of the savanna leads to high herd mortality in bad years and corresponding hardship and famines among the nomadic population who have increased in size and are dependent on the larger herds. This is followed by outside shipments of food, usually to a central location where the nomads and their cattle assemble, further contributing to overgrazing and erosion. Efforts aimed at aiding pastoral people by digging artificial wells and pumping underground water have resulted in more sedentarism and more range degradation. The latter is a direct result of an effort to divert more energy and materials from the system (smaller ranges and larger herds) without a corresponding change in the level of knowledge and investment.

Nomadic pastoralists are blamed for this situation brought about mostly by forces outside their societies. However, nomadic pastoralism if not interfered with is a successful management technique for the very dry savannas (rainfall below 600 mm) where rain-fed agriculture is not possible. Whether it can survive within the context of the modern African state is questionable. The lifestyle of present-day pastoralists is threatened by conflicts with agriculturalists and city dwellers whose vision of life is at odds with that of the herders. Furthermore, pastoralism can sustain only a certain human population density. This is now exceeded in many regions of Africa.

Some people claim that nomadic pastoralists have no appreciation of the ecological integrity of the savanna (Lamprey 1983). African pastoralists certainly do not have the same awareness of the ecological mechanisms underlying the relation between vegetation, climate and grazing that modern science possesses. Yet their traditional techniques have allowed their survival over millennia in a very harsh environment. Nomadic pastoralism in its traditional application probably no longer suffices to meet the needs of an expanding African human population. Neither can it meet the aspirations of the pastoralist population for a more sedentary and material lifestyle. It is, however, a sustainable system, if not interfered with, from which undoubtedly much can be learned.

SAVANNA USES INVOLVING SOME INPUTS

Between these traditional extensive grazing and cropping systems and the intensive high-input agricultural strategies described in the next section, a large number of management systems aimed at improving yields with a minimum of inputs is found.

In dry savanna regions (rainfall below 600 mm) agriculture is impracticable due to the small quantity and irregular distribution of rainfall; in very wet regions (rainfall above 2000 mm) epidemics and predators prevent pastoral development. In intermediate areas both stock-raising and agriculture are possible. Yet, without considerable inputs of capital and energy, agricultural yields are very low. Extensive ranching is consequently the most frequent type of modern utilisation of the seasonal savannas. Its techniques do not differ fundamentally from those of nomadic pastoralism, since they involve utilisation of the natural or modified vegetation, extensive use of the resources of the savanna and the organisation of animals into herds. Modern ranching has not proven to be as sustainable as traditional herding. It requires greater input of capital and knowledge, but considerably less of labour than pastoralism. Consequently, although individual animal productivity is higher, herd and human densities are lower.

Ranching may be defined as a form of livestock management in which the herds are normally kept for sale in a specific, sometimes fenced, area. Herds are divided into different age and sex groups, parasites and predators are checked, and the range is managed by controlling the use of water and fire. The technical differences between nomadic pastoralism and ranching are mainly quantitative, but there is a fundamental socio-economic difference. Ranching produces animals exclusively for a market; there is not the parallel reproduction of herds and of human groups that characterises nomadic pastoral societies. More generally, ranching has a different economic rationale and other forms of working relations, principally salaried, are dominant. This results in a considerable increase in the productivity of labour. Land tenure considerations also differ significantly between ranching and nomadic pastoralism. Thus the real division between nomadic pastoralism and ranching is a socio-economic one.

Pastoralists and ranchers have different management objectives. The aim of pastoralists is to maintain a large herd as an insurance against extreme drought or hardship. The goal of the rancher is to maximise money profit. Pastoralists try to reduce their workload as much as possible, ranchers try to maximise the productivity of labour. Their different perceptions of land use can lead to serious conflicts, as explained by Mentis and Seijas in their chapter.

Ranching encourages the rationalisation of savanna land use on a scale not known to pastoralists. However, the disorganised and competitive development of ranching enterprises, and the quest for short-term maximisation of profits leading to inadequate investments, can have undesirable consequences, such as soil degradation from overuse, disappearance of wildlife, destruction of vegetation by the misuse of fire, invasion by unpalatable plant species, and wind and water erosion resulting from overgrazing. These adverse effects can be seen in all tropical savannas of the world where traditional management has been replaced by extensive, market-oriented ranching (see chapter by Holmes and Mott). This has led in some quarters to the belief that the optimal sustainable way to utilise tropical savannas is to transform them into artificial pastures or agricultural fields (Goedert 1990). This belief is being increasingly challenged.

Principles of management of native pastures

Savannas may contain 100 or more species of grasses and forbs, but usually fewer than 10 provide most of the standing biomass and productivity. Forbs, especially species of *Leguminosae*, may supply very high-quality fodder but are usually not present in sufficient abundance to become an important component of the animal diet. Cattle graze preferentially the most palatable species, and if allowed to roam freely tend to reduce significantly the abundance of desirable species, allowing the less palatable to increase. Furthermore, the quality and quantity of fodder change dramatically throughout the year. At the beginning of the rainy season, the new expanding leaves provide the best fodder. As the wet season progresses, the quantity of grass increases but its quality decreases, with an increase in the proportion of silica and hardened tissues in the fodder. With the onset of the dry season, aerial biomass starts to dry out and decompose, and both the quantity and quality of fodder decrease dramatically.

Extensive ranchers do not tend their livestock as pastoralists do. Animals are allowed to roam freely. Instead of the tame animals characteristic of pastoralist herds, animals in extensive ranches are in an almost feral state (see chapter by Holmes and Mott). This can result in injuries and loss of calves.

To succeed in ranching on natural savanna pastures, the manager of an enterprise must determine the relative merit of the different grass species for animal production, vegetation stability and soil protection. One management objective must be to keep this optimal range structure. Changes in plant species composition and in species proportions over time are important indicators of the deterioration of the natural pasture. Another problem experienced by tropical savanna ranchers especially is continuous change in the floristic composition with an increase in unpalatable species. Continuous grazing leads to a reduction in the productivity and standing biomass of palatable vegetation. Soil erosion, surface crusting and reduced infiltration often accompany the loss of desirable perennial grasses through continual overgrazing. The time-scale involved in pasture change is often very long. Animal productivity in the short term is a poor guide to pasture condition.

Fire is used widely in the management of native savanna grasslands. Fire removes unpalatable dry material and, if applied properly and at the appropriate time, promotes grass growth. Fire, however, also destroys the seeds and seedlings of grasses and woody plants. The position of the perennating buds of grasses and the regime of reproduction determine the precise effect of fire on the regeneration of each species. Continuous use of fire therefore can also result in changes in species composition. Fire also leads to loss of the mineral matter contained in the vegetation, especially if applied early in the dry season or during the growing season. Timing, intensity and frequency of fire affect the physical environment and the biota.

One way to overcome the problem of the poor nutritional quality of native grasses is to oversow the savanna with more palatable species combined with a treatment of fertiliser, especially phosphate dressing. The following account of this technique as practised in the dry savannas of northern Australia (Queensland and Northern Territory) is based largely on the work of Edye and Gillard (1985).

The method consists of sowing directly into the pasture seeds of exotic species such as *Stylosanthes* spp., a leguminous species originally from South America but improved in Australia, or *Leucaena* spp. This method has the advantage over total replacement of the savanna with a sown pasture of lower costs as a result of not having to remove existing vegetation, both herbaceous and woody. Aircraft can be used to broadcast the seed and the fertiliser, further reducing costs. The density of seed applied is at the rate of 1 kg/ha. The establishment of the oversown seed is often slow due to competition by the native grasses. Burning, cutting or intense grazing of the savanna before sowing hastens establishment. It is recommended that the sown pasture be grazed lightly during the first year. The plants can thus become well established and have an opportunity to seed copiously during the first year. In Australia the rate of establishment of *Stylosanthes* is high because of the lack of native pests and parasites and the ability of legumes to fix nitrogen. Oversown pastures permit a doubling of the stocking rate and result in substantial increases in the growth rate of steers. Superphosphate fertiliser treatment can be applied at the time of treatment or after the legumes have become established, at a rate of about 220 kg/ha.

Improved pastures, either by the amelioration of the quality of the forage or the addition of watering points, are still subject to periodic droughts which will reduce the availability of fodder. Efficient management demands that stocking rates be decreased in order to avoid overgrazing, soil erosion and permanent range degradation. Ranchers are reluctant to do this. Economic considerations and the expectation that the drought will be short tempt them to delay this decision until the range is overgrazed and the animals begin to lose weight and decline in value.

Generally, the higher the initial stocking rate the greater the danger of permanent damage during drought (Mott 1986; Pressland and McKeon 1989). This is a further example of the need for greater investment (in this case prudent management) when humans divert more energy and materials from the ecosystem to themselves.

Forest production

Management of natural savannas may also be directed toward forest production, with the objective of encouraging the growth of trees. Savanna trees provide fuel for cooking or charcoal-making in much of the New and Old World tropics. In areas of high human density, such as regions of Africa, savannas have lost much of their trees. In the *cerrado* region of Brazil, charcoal production for the steel industry has resulted in the elimination of most savanna trees over large regions.

INTENSIVE, HIGH-INPUT MANAGEMENT

Most intensively managed savanna grasslands are created by the removal of the natural vegetation (especially the woody component) and replacement with either an artificially sown pasture or, more often, a crop. Soil texture, soil water and nutrient availability remain the limiting factors, although pest and disease problems further limit production. Without grazing, cutting or fire these savannas revert to their previous state. That is, these intensively managed grasslands are ecologically more unstable, and they require careful management to avoid changes in botanical composition in response to grazing or cutting regimes and fertiliser use (Snaydon 1980, 1984). Failure to recognise this can result in loss of what can be a substantial investment.

The establishment of artificially sown pastures or agricultural fields requires considerable inputs, supporting services and managerial skills. Although sown pastures are much more productive than natural savanna grasslands, their economic justification needs careful consideration. In the absence of irrigation, sown pastures provide ample fodder only during the rainy season. As the pasture goes dormant during the dry season, a shortage of fodder is experienced. Sown pastures must also be managed much more carefully to avoid quick deterioration and loss of investment.

Constraints on agriculture in tropical savannas

Where rainfall exceeds 600 mm native savannas are being replaced slowly but inexorably by agriculture to feed the expanding tropical population, especially urban dwellers, and as a source of foreign exchange for savanna countries (see chapter by Klink, Moreira and Solbrig). Agriculture in tropical savannas is limited by poor soil texture and fertility and highly seasonal rainfall. Weeds and pests constitute additional problems. On the other hand, light and temperature conditions are very favourable for cultivation.

High-input agriculture based on the production of cash crops is possible only if there exists an infrastructure for the transportation and marketing of the crop. It also requires financial and credit institutions, agricultural research organisations and institutions for the diffusion of agronomic knowledge. Very often these institutions are lacking or are deficient, and financing and inadequate agronomic techniques are imported from other regions. This often has disastrous economic and ecological consequences.

The low fertility of most tropical soils requires extensive soil preparation before intensive agriculture is possible. The major problems are very low pH, and a low concentration of phosphorus and nitrogen. These problems can be solved by applying lime and phosphate rock prior to cultivation (Goedert 1990). Nitrogen fertiliser is applied yearly. An application renders the field manageable for 10–20 years without additional treatment. In many savanna soils there are also deficiencies of micronutrients, such as sulphur or boron in many Brazilian soils. These can be controlled by yearly fertiliser application. Whether high-input agriculture is profitable depends on the costs of the inputs and the price of the crop. Soybean and rice have proved profitable in Brazil (see chapter by Klink, Moreira and Solbrig). High-input agriculture has obtained direct and indirect subsidies from governments (including Brazil) eager to obtain agricultural independence or to earn foreign exchange. It is not clear whether tropical agriculture can be made profitable without subsidies with today's commodity prices.

High-input agriculture and artificial pastures create the greatest transformation of the natural environment. Graded land results in changed run-off patterns. Native vegetation is totally eliminated, especially the woody component, increasing the danger of wind and water erosion. In many areas, such as in Brazil (see chapter by Klink et al.), chemical fertilisers, herbicides and pesticides have leached into underground aquifers used as a source of drinking water, previously endangered by high microbial counts. Poor or non-existent quality controls on potable water supplies worsen the problem. Roads, buildings and other infrastructure often interfere with natural run-off and create erosion. Finally, high-input agriculture tends to displace pastoralists and native subsistence agriculturists from the best soils. Pastoralism and native agriculture have persisted in these regions, sometimes for thousands of years, because of the use of low-yielding but sustainable techniques. Displacing native agriculture to marginal areas increases the danger of soil erosion in the regions and can create food supply problems for these people.

CONCLUSION

Generally in the literature, there are policy studies dealing with savanna use from an economic point of view and there are agronomic studies dealing with plant growth from a biological point of view. These two approaches need to be integrated for a realistic description of the ecosystem and to develop more sustainable management techniques. As was pointed out in the introductory chapter by Solbrig and Young, this division stems from a dual vision of nature, one that considers humans as outsiders looking into nature and extracting resources from it. Rather, we believe that people are another species in the system. Because of our large numbers and require-ments and our great competitive ability, humans may be endangering the functioning of the entire system, including our own survival (which obviously depends on the harmonious functioning of the whole).

Larger populations, higher incomes and increased consumption of animal protein around the world have intensified the pressure on land use, including savanna lands. Malthusian and neo-Malthusian theories maintain that population growth will reduce food surpluses and cause eventual starvation of tropical populations. This has created the fear that population pressure will lead to the massive degradation of savanna lands. Supposed desert-ification in the Sahel and other dry and semi-arid regions is used as evidence.

Not everybody agrees with this assessment. The Danish agricultural economist Esther Boserup (1981) has proposed that throughout history population increase has been the driving force promoting agricultural innovation. According to Boserup, population growth motivates people and facilitates technological change. Consequently, she feels that population growth can have a positive effect on development. Furthermore, people can adapt to increases in number by means other than the intensification of agriculture (Boserup 1985). They can change their diet from plants that are inefficient in the conversion of light energy to more efficient ones (from wheat to corn, for example), or to crops, with higher per unit surface yields (from seed to root crops for example). They can use their larger numbers to conquer neighbours with food surpluses and extract tribute, or they can trade for food.

Tropical savannas are among the last regions being converted from

42

extensive, low yielding, but generally stable, agro-pastoral use to more intensive employment. It is not possible to state categorically that the shift is driven by the massive population increases of tropical human populations, but it is very likely. Population increase is however not the only force impelling these changes. The drive for modernisation inherited from the recent colonialist past is a major factor of change in Africa (Kituyi 1990). Yet it is undeniable that intensification has introduced new agricultural and animal husbandry technologies. Some of these technologies have been borrowed, certain others modified, and some invented. A great deal of agricultural research and experimentation has been and is still going on.

During the past decades, there has been a significant increase in agricultural production due to technological advances (Wolman and Fournier 1987). While the new technologies have led to an increase in input factors (i.e. capital, energy, water, fertiliser and pesticides), they have also brought about environmental deterioration, such as soil erosion and reduction in soil fertility, groundwater contamination, air pollution and deforestation. The environmental impact of modern high-input agricultural practices is the most critical issue in the search for a high-yielding but sustainable agricultural technology.

To explore the interactions of resources, technology and environment for policy purposes requires careful analysis to quantify:

(1) How soil and climate determine appropriate technologies;

(2) How production technologies affect soil and water resources; and

(3) How to select sound production technologies that meet economic objectives and environmental goals.

In an insightful study of sustainable development in agriculture, Parikh (1988) pointed out that there has been a shift in perception of the problems of sustainability of agriculture, from a preoccupation with resource shortages to concern with environmental impacts. No doubt the great success of modern agriculture in meeting the food needs of the world's growing population has been largely responsible for this shift. However, as pointed out by Parikh, the problem of the sustainability of different types of land use demands that attention be paid simultaneously to production potentials and environmental consequences in searching for adequate technologies.

A very crucial issue is whether to continue developing management techniques based on hypotheses that assume that ecosystems are in ecological equilibrium or whether to abandon that paradigm in favour of non-equilibrium hypotheses. The ecological community is increasingly favouring the latter approach. At the minimum, we recommend that greater attention be paid to the dynamic behaviour of managed savannas. If savannas are equilibrium systems, then any alteration is almost by definition a degradation of the system and should be minimised. If, as seems increasingly more likely, savannas, especially dry savannas, are non-equilibrium systems, then changes are as likely to be improvements as deteriorations. Non-equilibrium systems will require more rather than less human investment to understand and manage them.

Edward O. Wilson of Harvard University has coined the very felicitous phrase the "stewardship of nature" to refer to the management of natural landscapes. In the European Middle Ages the steward was the person in charge of the affairs of the manor or castle. He saw to it that land and resources were apportioned fairly and according to ancient and accepted usage. He saw to the interests of his master, the lord of the manor, but he also was concerned with the welfare of the villeins. There were inevitable conflicts between the various interests and he tried to resolve them to the best of his abilities. If he was a good steward then the fields would produce enough for everybody, the barns and larders would be full in the winter, the cattle fattened and in condition to survive the cold season, and enough fodder stored. Of course there were good and bad stewards, and many indifferent ones too.

We all now have to become the stewards of ecosystems and allocate the resources between the need of our masters – human societies and human needs – and the needs of the "natural" part of the ecosystem. In medieval society the lord of the manor or castle could not survive without the produce of the villeins who took care of the fields. In our present society we will not be able to survive unless we allow plant and animal communities to function. The challenge is to understand how they function.

ACKNOWLEDGEMENTS

I would like to acknowledge Carlos Klink, Adriana Moreira and John Mott for a careful and critical reading of the manuscript and many helpful suggestions. I also wish to acknowledge a grant in aid of research from the US Department of Agriculture.

REFERENCES

Abreu, J. Capistrano de, (1954) *Capitulos de Historia Colonial 1500–1800*. Sociedade Capistrano de Abreu, Rio de Janeiro

Avilan, R.L. and Rojas I. de (1975) Evaluación de los niveles de azufre en suelos de la serie Barinas. *Agronomia Tropical* 25:149–61.

Azara, F. (1954) *Memoria sobre el estado rural del Rio de la Plata y otros informes*. Editorial Bajel, Buenos Aires

Bailey, H.P. (1966) The mean annual range and standard deviation as measures of dispersion of temperature around the annual mean. *Geografiska Annaler* 48A:183–94.

Bailey, H.P., Simpson, B.B., and Vervoorst, F. (1977) The physical environment: The independent variable. In: G. Orians and O.T. Solbrig (eds) *Convergent evolution in warm deserts.* Dowden, Hutchinson & Ross, Stroudsburg, Pennsylvania, pp. 13–49.

Boserup, E. (1981) *Population and technological change. A study of long-term trends.* University of Chicago Press, Chicago.

Boserup, E. (1985) The impact of scarcity and plenty. In: I. Rotberg and T.K. Rabb (eds) *Hunger and history,* Cambridge University Press, Cambridge, pp. 185–209.

Bourlière, F. (1983) *Tropical Savannas.* Ecosystems of the World vol. 13. Elsevier, Amsterdam.

Chevalier, F. (1975) *La formación de los latifundios en México.* Fondo de Cultura Económic, México.

Cochrane, T.T., de Azevedo, L.G., Thomas, D., Madeira Netto, J., Adámoli, J. and Verdesio, J.J. (1985) Land use and productive potential of American savannas. In: J.C. Tothill and J.J. Mott (eds) *Ecology and management of the world's savannas.* Australian Academy of Sciences, Canberra, pp. 114–24.

Coutinho, L.M. (1982) Ecological effects of fire in Brazilian cerrado. In: B.J. Huntley and B.H. Walker (eds) *Ecology of Tropical Savannas.* Springer, Berlin, pp. 273–291.

DeAngelis, D.L. and Waterhouse, J.C. (1987) Equilibrium and nonequilibrium concepts in ecological models. *Ecological Monographs* 57:1–21.

Edye, L.A. and Gillard, P. (1985) Pasture improvement in semi-arid tropical savannas: A practical example in Northern Queensland. In J. Tothill and J. Mott (eds) *Ecology and management of the world's savannas.* Australian Academy of Sciences, Canberra, pp. 303–9.

Ellis, J.E. and Swift, D.M. (1988) Stability of African pastoral ecosystems: alternate paradigms and implications for development. *Journal of Range Management* 41:450–59.

Frost, P., Menaut, J.C., Medina, E., Solbrig, O.T., Swift, M. and Walker, B. (1985) Responses of savannas to stress and disturbance. *Biology International,* Special Issue No. 10.

Frost, P.G.H. and Robertson, F. (1987) The ecological effects of fire in savannas. In: B.H. Walker (ed) *Determinants of tropical savannas.* IUBS, Paris, pp. 93–140.

Gillon, D. (1983) The fire problem in tropical savannas. In: Bourlière F. (ed) *Tropical savannas.* Elsevier, Amsterdam, pp.617–41.

Goedert, W. (1986) *Solos dos cerrados. Tecnolgias e estratégias de manejo.* EMBRAPA. Planaltina, Brasil.

Goedert, W. (1990) Estrategias de manejo das savanas. In: G. Sarmiento (ed), *Las sabanas Americanas.* CIELAT, Merida, pp. 191–218.

Gualdron, R. and Salinas, J. (1982) El azufre en suelos de los llanos orientales de Colombia. *Suelos Ecuatoriales, Colombia* 12:221–30.

Guerrero, R. (1971) Soils of the Colombian llanos orientales. Composition and classification of selected profiles. PhD thesis, North Carolina State University. 78 pp.

Guerrero, R. and Cortes, A. (1976) Caracterización y clasificación de perfiles seleccionados de suelos de C.N.I.A. La Libertad y zonas aldeañas. Instituto Colombiano Agropecuario. Instituto Geográfico Agustin Codazzi, *Bol. de Investigacion* 146:131 pp.

Hawkes, J.G. (1983) *The diversity of crop plants.* Harvard University Press, Cambridge.

Holling, C.S. (1973) Resilience and stability of ecological systems. *Annual Review of Ecological Systems* 4:1–23.

Hudson, N.W. (1975) *Soil conservation.* Batsford, London.

Huntley, B.J. and Walker, B.H. (1982) *Ecology of tropical savannas.* Springer, Berlin.

Jones, M.J. and Wild, A. (1975) Soils of the West African savanna. *Commonwealth Agricultural Bureaux (CAB) Technical Communication* 55:1–246.

Kituyi, M. (1990) *Becoming Kenyans. Socio-economic transformation of the pastoral Maasai.* ACTS Press, Nairobi, Kenya.

Komarek, E.V. (1972) Lightning the fire ecology in Africa. *Proceedings of the Tall Timbers Fire Ecology Conference* 11:473–509.

Lamprey, H.F. (1983) Pastoralism yesterday and today: the overgrazing problem. In: F. Bourlière (ed) *Tropical savannas.* Elsevier, Amsterdam, pp. 643–66.

Leon, J.C. and Botero, P.J. (1980) *Geología, geomorfologia y suelos de la region de la cordillera oriental, Villavicencio, Puerto Lopez, Puerto Gaitan.* Publicación interna, CIAF.

Little, M. and Leslie, P.W. (1990) The South Turkana Ecosystem Project. Report to the

Government of Kenya, Office of the President. N.Y. Binghampton, Department of Anthropology, State University of New York.

Lopes, A.S. and Cox F.R. (1977) A survey of the fertility status of surface soils under "cerrado" vegetation in Brazil. *Journal of the Soil Science Society of America* 41:742–47.

Medina, E. (1978) Significación ecofisiológica del contenido foliar de nutrientes y el área foliar específica en ecosistemas tropicales. In: *Actas del III Congreso Latinoamericano de Botánica.* Brasilia.

Medina, E. (1987) Requirements, conservation and cycles of nutrients in the herbaceous layer. In: B.H. Walker (ed) *Determinants of tropical savannas.* IUBS, Paris, pp. 39–65.

Menaut, J.C., Barbault, R., Lavelle, P. and Lepage, M. (1985) African savannas: biological systems of humification and mineralisation. In: J.C. Tothill and J.J. Mott (eds) *Ecology and Management of the World's Savannas.* Australian Academy of Sciences, Canberra, pp. 14–33.

Montgomery, R.F. and Askew, G.P. (1983) Soils of tropical savannas. In: F. Bourlière (ed) *Tropical savannas,* Elsevier, Amsterdam, pp. 63–78.

Mott, J. (1986) Planned invasions of Australian tropical savannas. In: R.H. Groves and J.J. Burdon (eds) *Ecology of Biological Invasions: An Australian Perspective.* Australian Academy of Science, Canberra, pp. 89–96.

Mott, J.J., Williams, J., Andrew, M.H. and Gillison, A.N. (1985) Australian savanna ecosystems. In: J.C. Tothill and J.J. Mott (eds) *Ecology and management of the world's savannas.* Australian Academy of Sciences, Canberra, pp. 56–82.

Nicolis, G. (1991) Non-linear dynamics, self-organisation and biological complexity. In: O.T. Solbrig and G. Nicolis (eds) *Perspectives on biological complexity.* IUBS, Paris, pp. 7–49.

Nicolis, G. and Prigogine, I. (1977) *Self-organisation in non-equilibrium systems.* Wiley, New York.

Nicolis, G. and Prigogine, I. (1989) *Exploring complexity.* Freeman, New York.

Noy-Meyer, I. (1975) Stability of grazing systems: an application of predator-prey graphs. *Journal of Ecology* 63:459–81.

Nye, P.H. and Greenland, D.J. (1960) *The soil under shifting cultivation.* Commonwealth Bureau of Soils, Harpendern, England.

Odegi-Awuondo, C. (1990) *Life in the balance: ecological sociology of Turkana nomads.* ACD Press, Nairobi.

Parikh, J.K. (ed.) (1988) *Sustainable development in agriculture.* Martinus Nijhoff Publishers, Dordrecht.

Pelissier, P. (1960) *Les paysans du Senegal. Les civilisations agraires du Cayor a la Casamance.* Fabregue, Saint-Yrieix [Haute-Vienne].

Pressland, A.J. and McKeon G.M. (1989) Monitoring Animal Numbers and Pasture Condition for Drought Administration – An Approach. Range Monitoring Workshop, 5th Soil Australian Conservation Conference. W.A. Agriculture Department, Perth.

Sanford, W.W. (1982) The effect of seasonal burning: A review. In: W.W. Sanford, H.M. Yefusu and J.S.O. Ayensu (eds) *Nigerian savanna.* Kainji Research Institute, New Bussa, pp. 160–188.

Santamaria, F. (1965) Distribución geográfica del arrecife en Venezuela. *Bol. Soc. Venezolana de Ciencias Naturales* 25:350–54.

Santamaria, F. and Bonazzi, A. (1963) Factores edáficos que contribuyen a la creación de un ambiente xerofítico. *Bol. Soc. Venezolana de Ciencias Naturales* 25:9–17.

Santamaria, F. and Bonazzi, A. (1964) Estudio sobre la permeabilidad del arrecife. *Bol. Soc. Venezolana de Ciencias Naturales* 107:175–86.

Sarmiento, G. (1983) The savannas of tropical America. In: F. Bourlière (ed) *Tropical savannas.* Elsevier, Amsterdam, pp. 245–288.

Sarmiento, G. (1984) *The ecology of neotropical savannas.* Harvard University Press, Cambridge.

Sarmiento, G. and Monasterio, M. (1971) Ecología de las sabanas de América tropical. I. Análisis macroecológico de los llanos de Calabozo, Venezuela. *Cuadernos Geográficos No 4. University Los Andes,* Mérida. 126 pp.

Silva, J.F. (1972) Influencia de los procesos pedogenéticos en la diferenciación de comunidades y en el comportamiento de las especies en los Llanos Occidentales de Venezuela. *Trabajo* de Ascenso, Facultad de Ciencias, Univ. Los Andes. Mérida, Venezuela.

Silva, J.F. (1987) Responses of savannas to stress and disturbance: species dynamics. In: B.H. Walker (ed) *Determinants of tropical savannas*. IUBS, Paris, pp. 141–156.

Silva, J.F. and Sarmiento, G. (1976a) La composición de las sabanas de Barinas en relación con las unidades edáficas. *Acta Científica Venezolana* 27:68–78.

Silva, J.F. and Sarmiento, G. (1976b) Influencia de los factores edáficos en la diferenciación de las sabanas. Análisis de Componentes Principales y su interpretación ecológica. *Acta Científica Venezolana* 27:141–47.

Snaydon, R.W. (1980) Plant demography in agricultural systems. In: O.T. Solbrig (ed) *Demography and evolution in plant populations*. Blackwell, Oxford, pp. 131–160.

Snaydon, R.W. (1984) Plant demography in an agricultural context. In: R. Dirzo and J. Sarukhan (eds) *Perspectives in plant population ecology*, Sinauer Associates, Sunderland, Massachusetts, pp. 389–407.

Stott, P. (1984) The savanna forests of mainland South East Asia: an ecological survey. *Progress in Physical Geography* 8:315–35.

Torres, C.L. (1980) *Genesis de los Suelos del Vichada*. Instituto Geográfico Agustin Codazzi, Bogotá.

UNESCO (1979) *Tropical grazing land ecosystems*. UNESCO, Paris.

Vargas, E. (1964) *El aluminio de cambio en los suelos de los llanos orientales*. IGAC, Bogotá.

Walter, H. (1969) El problema de la sabana. *Bol. Soc. Venezolana Ciencias Naturales* 28: 123–44.

Walker, B.H. (1985) Structure and Function of Savannas: An Overview. In: J.C. Tothill and J. Mott (eds) *Ecology and management of the world's savannas*. Australian Academy of Sciences, Canberra, pp. 83–91.

Westoby, M., Walker, B. and Noy-Meyer, I. (1989) Range management on the basis of a model which does not seek to establish equilibrium. *Journal of Arid Environments* 17: 235–39.

Wiens, J.A. (1977) On competition and variable environments. *American Science* 65:590–97.

Wiens, J.A. (1984) On understanding a non-equilibrium world: myth and reality in community patterns and processes. In: R. Strong, Jr., D. Simberloff, L.G. Abele and A.B. Thistle (eds) *Ecological communities: conceptual issues and evidence*. Princeton University Press, Princeton, New Jersey.

Wolman, M.G. and Fournier, F.G.A. (1987) *Land transformation in agriculture*. John Wiley & Sons, Chichester.

Young, A. (1976) *Tropical soils and soil survey*. Cambridge University Press, Cambridge.

Zinck, A. and Urriola, P. (1970) *Origen y evolución de la Formación Mesa, un enfoque edafológico*. Min. Obras Públicas, División de Edafología, Barcelona, Venezuela.

CHAPTER 3

LAND TENURE FOR PASTORAL COMMUNITIES

I. Scoones, C. Toulmin and C. Lane

SUMMARY

Communal grazing systems within Africa's savanna lands are under threat from various directions. Growing human populations put such systems under increased pressure, while inappropriate interventions, based on misconceptions regarding traditional land use management, have contributed towards a breakdown in communal tenure arrangements. The trend towards privatisation of the pastoral commons has brought mounting social, economic and environmental costs.

A consensus is now emerging which stresses the need to develop methods of protecting key resources, especially the watering points and strategic reserve grazing areas that are pivotal to the operation of entire grazing systems. A framework that strengthens local institutional management capacity needs to be established, and local management institutions must be provided with administrative support to assist them in ensuring that traditional rights remain enforceable. The legal framework that facilitates development of these institutions must be quite flexible and, at the local level, very negotiable.

Great care must be taken to ensure that social and economic policies encourage rational land use. A sense of resource ownership is critical for those who depend upon savanna resources for their survival. Providing that local management institutions are democratic and responsive to local, regional and national needs, devolution of management responsibility to local pastoral communities offers considerable advantages.

INTRODUCTION

Throughout the world, savanna areas are characterised by increasing resource pressure. Located between arid lands and higher potential forest and agricultural lands, the African savanna zone has had to absorb increasing numbers of people. Pastoralists from the drylands and agriculturalists from the subhumid zones are both expanding into these areas, as the traditional land available to them diminishes. The savanna zones are thus sites of increasing land-use conflicts, due to land-use changes, population pressure and insecurity brought about by war.

This chapter examines land-use conflict and land tenure issues in savanna areas, drawing upon a range of case studies on community-based pastoral grazing systems. It begins with an examination of the theoretical debate about land tenure by identifying three alternative conceptual frameworks, and then deals with the socio-political content of land-use conflicts in many savanna areas. It concludes with a discussion of various policy implications for effective land right allocation and land-use management in areas where community-based pastoral grazing systems still exist. Much of the material we draw on is from the African savannas. Similar processes are evident in other parts of the world.

Arid, semi-arid and sub-humid regions are estimated to make up 35 per cent of the Earth's surface, and contain one fifth of the world's population (UNEP 1984). Particular concern has been focussed on the risks of "desertification" faced by these regions. In 1983, it was estimated that 75 per cent of land in arid, semi-arid and subhumid regions had suffered some loss of productivity. In some cases this loss was estimated to be as great as 25 per cent (UNEP 1984). Later studies questioned both the definition of "desertification" and the basis on which these statistics were calculated (Warren and Agnew 1988; Nelson 1990). Little serious credence is now given to the claim that deserts advance and swallow up adjacent farmland.

Nevertheless, dryland regions throughout the world are being subjected to increasing pressure from the competing demands of different land users. The main focus of attention has now shifted from the desert margins to the higher potential subhumid savanna region, where human and livestock densities are much higher. Here, soil erosion and loss of both vegetative cover and soil fertility have caused increasing concern. Another problem is the decline in living standards and opportunities available to savanna populations.

In many parts of the world, common-property resources are under threat. Privatisation processes and land degradation have reduced the availability of common resources. These processes have greatest impact on the poor, who have few income sources and rely on wild resources, and marginal groups such as pastoralists who must rely on communal grazing lands to sustain their herds. Jodha (1986, 1991), for example, found that over 80 per cent of poor households in parts of the drylands of India relied on common lands for food, fuel, fodder and other vital products. Jodha also found that these common lands had shrunk by between 26 and 63 per cent over the past 30 years. Most of the privatised land ended up in the hands of richer segments of society. Studies within this book suggest that this has also occurred in Australia and South Africa, where indigenous peoples have not had equal opportunity to gain access to land; in Latin America, where land ownership is a sign of wealth; and in Kenya and Botswana, where private ranches are only allocated to relatively well-off pastoralists.

THE THEORETICAL DEBATE

The debate over appropriate forms of land tenure and management for savanna areas has increased in recent years. Three different approaches to resolving the problems faced by semi–nomadic pastoral communities have been adopted:

(1) *Privatisation regimes*, where individual pastoralists are granted exclusive rights to control access to specified areas of grazing land (Hardin 1968);

(2) *Common-property regimes*, where grazing access rights to large areas of land are managed collectively by communities of people who have close natural affiliations (Runge 1986); and

(3) *Co-ordinated access regimes*, where social institutions allocate access rights but enforcement of these rights is not dependent upon maintaining a sense of community (Swallow 1990; Lawry 1990).

Often the first regime is referred to as the "tragedy of the commons" model. Hardin (1968) argued that, when people have open-access rights to a resource, the private benefit of grazing cattle on a common pasture exceeds private cost. This is because all people using the pasture incur some of the cost, but the benefits accrue only to the individual. Thus open-access resources tend to be "over-grazed" in the sense that each person is encouraged to run more animals, as they expect to gain more than they lose. However, when a pasture is less resilient than the animals that graze it, resource degradation and even ecological collapse can occur.

Hardin therefore recommended forms of private or state–regulated ownership to avert such a tragedy. This model has dominated policy thinking for some time, and has been used to legitimise privatisation programmes and to support government intervention to ensure that environmental conservation objectives are achieved. The conceptual model represents a possible scenario in a true open-access situation, but is inappropriate for true common-property regimes, where individual use rights are conditioned by community obligations that seek to assure users that each other's collective interests will be respected (Ciriacy-Wantrup and

51

Bishop 1975; Runge 1981, 1986). Much empirical evidence now exists to dispute Hardin's proposition that the model describes what happens in many savanna areas (cf. Shepherd 1988). Customary savanna land use based on common-property resources has proved sustainable in the past.

Under most common-property regimes, community-based customs and rules define rights of access, and institutions exist to enforce them. Moreover, where productivity per unit area is low and the costs of establishing and enforcing clear property rights are very high, common-property regimes are likely to be more efficient than all other forms of tenure and may enable benefits to be distributed more equitably. Joint use may also be more beneficial when resource availability is variable in space and time. Interdependence in resource use may also help to forge community ties and thus help to hedge against uncertainty (Runge 1986).

In many situations, tight common-property regulation probably never existed; in such cases more flexible, informal and opportunistic regulatory systems may be more appropriate than ones based upon formal written law. Tenure systems – based on flexible, negotiated contracts and agreements seek to co-ordinate access rights, and may be the most appropriate form of land tenure for many savanna areas. Indeed, the formal models proposed for common-property management are not necessarily supported by empirical data (Swallow 1990; Lawry 1990). This tends to be the case when open economic systems encourage commercialisation, undermining the structures that maintain community interdependence and reducing incentives for community-level management.

Each of these theoretical models makes different assumptions about savanna land use and proposes different tenure regimes. The policy implications of these regimes – appropriate institutions, legal frameworks, incentives for land management etc. – differ considerably. It is important, however, to use empirical field-based evidence in assessing the relative merits of the models and in making appropriate policy recommendations. It is also important to recognise the historical and political contexts within which such decisions are taken (Kituyi, this book).

FORMS OF CONFLICT OVER SAVANNA LAND USE

Land-use conflicts in the African savannas are complex, and are due to both internal regional factors and external influences. Within the areas that still contain community-based grazing systems, conflicts exist between peasant agriculture and pastoralism; pastoralists and mechanised agriculture; pastoralists, cultivators and wildlife managers; with cultivating herders; and with absentee herd owners. Although many of these resource-use

52

conflicts can co-exist, we will deal with each area of conflict separately, for clarity.

Peasant agriculture and pastoralism

Land-use conflicts between peasant agriculturalists and pastoralist groups are perhaps the most widespread and intense form of conflict in savanna lands. Many examples could be given, including disputes over *fadama* lands between Hausa agriculturalists and Fulani pastoralists in northern Nigeria (Cline-Cole 1988), disputes in the Senegal river valley (Horowitz and Salem-Murdock 1991) and conflicts over grazing land in Maasailand (Parkipuny 1991).

Such disputes are often focussed on land that is vital for pastoral production and of greatest potential for agricultural production. Dry-season grazing land, for instance, is often high quality cropping land. Agricultural expansion results in encroachment on such areas, removing vital land from pastoral production. In many parts of the Sahel this has caused permanent loss of dry-season pastures and abandonment of the grazing transhumance (Schoonmaker-Freudenberger 1991, Touré 1990). This has had damaging consequences for the pastoral economy, as pastoral production is now reliant on arid annual grassland pastures and a diminishing number of dry-season refuges (Lane and Scoones, this book).

But options may exist for integrating pastoral and agricultural resource use, or negotiating effective land access rights. In central Mali, for example, herders pass many months of the dry season camped on farmland, manuring fields in exchange for access to village wells. Similar manuring contracts were formerly common throughout the West African savanna. This practice is now dying out, however, as farmers increasingly invest in their own cattle and the quantity of pasture available for grazing decreases because of agricultural expansion (Toulmin 1983).

Pastoralists and mechanised agriculture

Similar land-use conflicts are evident in areas where mechanised agricultural schemes have resulted in encroachment on pastoral areas. They have been documented in eastern Sudan (Sorbo 1985), central Sudan (Manger 1987) and northern Tanzania (Lane 1990; Lane and Pretty 1990).

Agricultural planners assumed that land was "underutilised" and could be more effectively exploited through mechanised agriculture. But experience has shown that State and privately run mechanised agriculture schemes established in the African savannas during the 1960s and 1970s have not been effective. Planners overestimated returns from mechanised schemes and underestimated their cost to pastoral systems (Lane and Scoones, this book).

More thorough economic and ecological planning is required if management strategies for the African savannas are to be effective. The full economic costs and benefits of the various land uses must be taken into account, and the different land users must be involved in the planning process.

Pastoralists, cultivators and wildlife managers

Large wildlife estates were established in many savanna areas of eastern and southern Africa during the colonial era. The consequent reduction in land available for pastoral or agricultural production resulted in serious conflicts of interest. State wildlife authorities argue for maintenance of exclusive wildlife areas as a source of tourist-derived foreign exchange. They claim that both the wildlife and the scenic values would inevitably be reduced if the land were made available for agricultural or pastoral use. They are supported by the international conservation lobby, which argues volubly for conservation of the wildlife heritage of the African savannas.

These conflicts have been documented in the Ngorongoro area of Tanzania (Arhem 1985), Maasailand in Kenya (Collett 1987) and the Zambezi valley of Zimbabwe and Zambia (Abel and Blaikie 1986). The West African region, with its fewer wildlife resources, has experienced much less alienation of pastoral lands for conservation purposes.

There is increasing evidence of attempts to offset elements of such conflict. Multiple-use arrangements for pastoral grazing in wildlife zones in Kenya have been attempted (Western 1982). The communal area management programme for indigenous resources (CAMPFIRE) programme in Zimbabwe, for example, aims to increase the flow of money from national parks and safari areas to local communities, through district level management committees. These committees have an overriding responsibility for controlling development within an area and redistributing the revenue and related benefits from tourism, hunting and wildlife culling (Zimbabwe 1986; Murphree and Cumming, this book). None of the institutional and financial arrangements for co-management of communal resources are perfect, however, and their improvement still remains a challenge.

Cultivating herders

Processes operating within pastoral societies may result in conflicts over land use. Little (1987), for example, documents land acquisition for cultivation by 11 Chamus pastoralists in the Baringo district of Kenya. Richer pastoralists are able to invest in irrigated plots while poorer pastoralists, with insufficient animals to sustain them, must resort to dryland farming. A similar case of internal differentiation is observed in Maasailand, where richer pastoralists have acquired common-grazing land for extensive cultivation, in anticipation of receiving private title to that land (Kituyi, this book; Grandin 1988).

Increases in cultivation in pastoral areas may occur through internal processes of social differentiation (Little 1987) or temporal adaptation to changing climatic circumstance (Anderson 1987), but often cultivation and sedentarisation are encouraged by state intervention. As the Kenyan example above illustrates, sedentarisation can occur in expectation of state intervention. Thus a programme in one area can have knock-on effects in other areas.

Settlement schemes for pastoralists have been attempted in many areas in response to a variety of objectives. In some cases, herders have been displaced from their traditional grazing lands and resettled in irrigated agriculture (Sorbo 1985). Elsewhere, settlement has been encouraged to improve access to services, such as health and education. Rarely, however, have such services been adapted to suit mobile populations (Swift et al. 1989; Ndagala 1982). Settlement of nomadic populations has also been pursued for political reasons, with governments particularly eager to acquire some measure of political control over mobile communities.

In many areas, it is increasingly inappropriate to view conflict simply in terms of herder-farmer group disputes. Social differentiation within agricultural and pastoral groups results in some farmers holding large livestock herds and some pastoralists increasingly concentrating on agricultural production. As livelihood strategies change, so do land claims and appropriate forms of land tenure. Those who lose out are those with less political or economic influence. In the early years of group ranch policies in Kenya, for example, there was widespread support for the idea. Group ranches were seen as one way for the Maasai to affirm rights over grazing land in the face of increasing encroachment by neighbouring farmers (Oxby 1981; Little 1985). Group ranches, however, soon became a vehicle for the powerful Maasai (and others) to acquire and effectively privatise land, to the exclusion of others (Grandin 1988). This dynamic has been observed in other attempts at establishing group grazing management systems in Botswana (Lawry 1990; Sweet 1987) and in Zimbabwe (Cousins 1988). Thus, attempts at initiating group management schemes in savanna areas must be explored in the light of the ongoing dynamics of social and political differentiation.

Absentee herd owners

Livestock on savanna lands is not necessarily owned by those people who manage it. Agriculturalists, traders, politicians and businessmen have increasingly invested in livestock, resulting in a pattern of absentee herd ownership. The political dynamics of herd ownership are significant for land-tenure and land-use conflict issues. Little (1985) documented such a process in Baringo, Kenya, where this phenomenon is increasing. New owners of livestock, with political or economic influence, are sometimes able to acquire land and privatise it for commercial grazing. Such "part-time pastoralists" operate outside the realm of control of local savanna land-management systems. In such cases the options are either the development of access-coordination frameworks within which common-property, private property and selective access arrangements can co-exist, or full privatisation. Such frameworks need to be able to cope with pastoralists who start cultivating, absentee landlords and so forth. Importantly, they must anticipate change, and facilitate its evolution in socially efficient and equitable directions.

SYSTEMS OF LAND MANAGEMENT: SOME CASE STUDIES

Recognising the nature of the above conflicts, it is useful to examine several case studies of communal land management in savanna areas of Africa.

Indigenous management in the Niger river delta, Mali

The nineteenth century Dina code established use and property rights over forage, agricultural and fisheries resources in the Niger delta. The regulations were enforced and sanctioned by the Fulani state. Grazing rights were allocated to clan groups of herders in a way that stipulated dates for grazing in the delta, transhumance routes and relationships between groups (Moorehead 1989; Lawry 1990). A code was established in order to reduce conflicts, regulate access and formalise the power of the State (including taxation rights). But over the last 50 years the system's effectiveness has decreased. The influence of the national government has increased and local social differentiation has undermined the ties and social bonds that maintained respect for the code. However, the traditional system could still form a basis for renegotiating rights and regulations for common-property management within the delta (Moorehead 1989).

Borana common-property regulation, Ethiopia and Kenya

Although there are few access restrictions to the Borana rangelands of Ethiopia and Kenya, the resource is regulated by the relatively limited number of wells that are spaced widely across this region (Helland 1982). In much of the Ethiopian rangelands, each well is controlled by a Borana clan and regulated by a council of well users. The few available wells and the slow, labour-intensive nature of water extraction limit the number of livestock that can be held, and so regulate rangeland use. In Kenya, however, well and irrigation projects have increasingly disrupted this form of regulation (Hogg 1987). The social technology has been unable to keep up with the material technology.

Range enclosure and land privatisation

Under conditions of heightened resource pressure, the incentive to privatise the range increases as people seek to maintain or improve their quality of life. Such pressures result from increased population, reduced rainfall and the land-use conflicts listed above. The effects of privatisation on those who lose access rights and do not receive exclusive rights to equivalent areas can be particularly adverse. In western Sudan, for example, sedentary livestock owners enclosed prime village grazing land, particularly during and following the major droughts of the mid-1980s. This effectively excluded transhumant pastoralists from using wadi grazing areas (Behnke 1985). Similarly, land tenure regulations in Somalia support enclosure of cultivated land. Following well construction, the surrounding land becomes increasingly valuable and some individuals are now endorsing both agriculture and grazing land. With the reduced amount of common land available, others have had no option but to follow this trend (Behnke 1988). Ultimately, the entire community-based property system breaks down.

Privatisation of high-value land may coincide with commercialisation of the production system. In some areas, State incentives for establishing land titles or for promoting beef production have encouraged this trend. The establishment of private ranches in Kenya's Maasailand, for example, has followed from the group ranch initiative and land-title registration occurring elsewhere in Kenya (Grandin 1988). Similarly, land was privatised under the Tribal Grazing Land Policy (TGLP) in Botswana. Typically, however, most of the land is held by a small class of wealthy and influential cattle owners (Sandford 1990).

Grazing schemes and group ranches

Forms of group management of grazing land have been the most popular policy advocated by donors and governments over the past 20 years. Examples include the TGLP in Botswana, the group ranches of Kenya, the grazing schemes of Zimbabwe and the grazing associations of eastern Senegal or Lesotho.

The success of group management policies has been mixed. Most of these projects have been successful in securing some grazing rights for pastoral producers in the face of encroachment by other grazing communities and aspiring agriculturalists. However, group rights have sometimes been usurped by powerful individuals, and land has become effectively privatised. It is unclear, therefore, whether the group management policies adopted during the last 2 decades have resulted in more effective savanna land management. Land users have often rejected regulations imposed by technical officers from State or donor agencies, because they are inappropriate. Regulations concerning fence lines, rotational grazing systems and stocking rates have been ignored and alternative, more locally appropriate, management systems have evolved (Grandin 1988; Scoones and Cousins 1991).

In addition, the level of cooperation required for effective group management has been lacking in some cases. Drawing on the experience of grazing associations in Lesotho, Lawry (1989) suggested that the reasons for the reduced cohesiveness of group management associations include: the multiple objectives of livestock owners; a lack of strong local institutions to make regulations effective; and options for generating alternative income from remittances. All these factors act to reduce commitment to communal management of local natural resources. In addition, the tendency for people to attempt to justify communal management schemes in terms of ecological objectives, rather than on the basis of direct economic gain to pastoral producers, has reduced the incentive for herders to co-operate in group management of these resources.

Village land management

In a number of Sahelian countries, village land-use management systems are evolving (known as "gestion de terroir"). Village-level committees are being given the responsibility and the right to plan how resources may be used within the village's territory and where community-wide investment should be made. For example, these committees may decide where to build soil and water conservation structures, the most appropriate areas for re-afforestation, and rules for regulating access to farmland, forests and grazing. At the same time, governments are devolving the power to control resources from centralised State structures to the local community.

Within pastoral areas, a parallel process is under way accompanied by considerable debate regarding how to devolve responsibility for pastoral resources to local herding groups. Many of the underlying problems have yet to be resolved. These include: defining the pastoral group with rights to manage this territory; and developing government mechanisms for enforcing the rules developed by management committees. One view, which accords with traditional management systems, is that water points should constitute the focus around which to define the pastoral territory given to a pastoral association (Thébaud 1990). Given the large water output of such bore-holes and their heavy maintenance costs, however, there can be immense financial problems associated with allocating public bore-holes to a given pastoral association. Processes of allocating rights to local communities also involve considerable legislative and practical difficulties.

Even where legislation or decrees grant rights to manage pastoral resources, it may be difficult to get local government to enforce these rules in practice, particularly if it means acting against the interests of powerful livestock owners.

LEGAL FRAMEWORKS FOR SAVANNA USE

A patchwork of different kinds of legislation has been stitched together to deal with land ownership and rights to different savanna resources. Colonial law, customary tenure and post-independence legislation govern the use of different kinds of resources and apply to different geographical areas, to varying extents. Resources of greatest value, such as irrigable land and land close to urban centres, have often been the focus of formal property claims, with firm title to those resources subsequently allocated by deed. In marginal farming and pastoral areas, however, people rarely have a piece of paper confirming their rights to use land. In cases of dispute, parties in conflict have recourse to customary and modern law, although final judgement often favours the party that can muster the greatest political and financial support.

The French-speaking countries of the West African Sahel make a distinction between rights to land acquired for cultivation purposes and that obtained for other purposes, such as grazing. "La mise en valeur", or putting land to good use, is the principle underlying governments' recognition of traditional rights to use land (Le Bris et al. 1991). Historically, pastoral herders have not been considered to be putting land to good use. Grazing is thought to be a transitory form of resource use, in which the passing herds effect no improvement to the resources they use. Consequently, pastoralists have increasingly considered it in their interests to start cultivating, at least in a rudimentary fashion, in order to acquire some land rights. This practice often results in the cultivation of unsuitable land, which removes ground cover and increases erosion risks.

Attempts at defining and enforcing tenure legislation result in a similarly perverse response to tree tenure in the Sahel (Elbow and Rochegude 1990). Highly centralised and regulatory forest codes are meant to prevent unlicensed cutting of listed tree species. Even trees planted by an individual cannot be cut by that person without a licence, which has to be bought from the local forestry service. While few people fully respect such legislation in practice, there is no doubt that these codes constitute a significant disincentive for people to plant and care for trees over which they will have only limited and ill-defined rights. Normally, a strong sense of ownership is a necessary prerequisite to sustainable investment.

Even where firm legislation exists on paper, it may still be very difficult for communities to ensure that their rights are fully respected. Urban-based commercial interests are often able to subvert official systems when those systems restrict their ability to earn high profits. A case study from the Bay Region of Somalia demonstrates the power of organised charcoal merchants, and their ability to intimidate local people who try to protect their forest resources from the charcoal burners. In practice, the charcoal merchants ignore official legislation, and as the legislation is not enforced this strategy succeeds (Shepherd 1988).

CONCLUSIONS: ISSUES FOR LAND TENURE POLICY IN SAVANNA AREAS

Many customary systems of savanna resource use have proved sustainable. Contrary to claims made in the "tragedy of the commons" model, these systems provide effective management of scarce and varied resources. Customary systems are therefore a good starting point for assessing different land-use options. Before new legal and administrative frameworks are imposed on savanna populations, greater consideration needs to be given to ways of enhancing local capacity to manage and protect resources effectively. In developing such systems, however, great care must be taken to ensure that they are capable of evolving with and adapting to the changing aspirations of many pastoral people. Increasing numbers of these people are becoming involved in non-grazing activities. Agriculture, politics, teaching and transport are but a few examples. In each case these activities challenge the old traditional regime that survived under a close-knit subsistence economy.

The earlier outlines of theoretical regimes for land management and empirical observations of land-tenure issues in the African savannas suggest a number of policy guidelines.

Identification of key resources

Certain areas within savanna landscapes are vital for production. In the context of pastoral production, these are often dry-season grazing areas and water points that provide key resources for local communal land management. It is around these areas that most land-use conflicts are centred, and where land privatisation is most common. Collective management is likely to be more successful if it is centred on high-value resources (Lawry 1990), and land of low productivity could also be more profitably managed collectively as it is too costly to privatise (Bromley and Cernea 1989). Identifying high-value key resources is thus essential for any effective land management strategy.

Local institutional management capacities

Effective land-use management must be based on local institutional management capabilities, as the imposition of alien management systems tends to fail. Codified, well-regulated and efficiently functioning common-property management systems can exist in some situations. In other settings, more flexible, opportunistic regimes are more appropriate. They may not constitute "management systems" as such, but they can nevertheless act effectively to regulate and manage common-property resources. Where local management is weak or ineffective, it may be necessary to restructure local institutions or to enhance their capacity before effective collective resource management is possible.

Processes of economic and political differentiation

Tenure systems do not operate in a political or economic vacuum. Individuals and organisations often wish to establish private ranches, wildlife reserves or tourist developments, for example, in savanna areas. These desires motivate them to exert influence on tenure systems, at both national and international levels. At the regional level, political factors and economic incentives often encourage conversion of open rangeland to private cultivation, resulting in agricultural expansion. At the local level, claims made by wealthy cultivating herders and absentee stock owners over

61

an area of land are very different to those made by traditional pastoralists (Little 1985). Recognition of the changing context within which tenure systems evolve is critical during policy formulation (Kituyi, this book).

Economic and political differentiation processes have a significant impact on the success of such systems. The appropriateness and effectiveness of any such systems are critically dependent on these processes. An understanding of the economic and political factors acting upon land tenure systems is therefore vital for effective savanna management.

Conflict resolution mechanisms

Processes for resolving land-use conflicts need to be established before effective savanna development is possible. The rights of different groups of land users to control and develop resources, especially those vital for production, need to be clarified. Mechanisms for negotiating access rights and land-use contracts (and for imposing sanctions when agreements are contravened) also need to be established. Land-use disputes will sometimes need to be adjudicated by an authority empowered for this purpose, and the legitimacy and impartiality of that authority must be guaranteed; land users tend to adopt less peaceful approaches to resolving land disputes when formal mechanisms are distinctly partisan.

Legal frameworks for land rights

The specification of clear, enforceable land-use rights is important in establishing effective land tenure systems in savanna areas. Such rights can be specified in enforceable leases, rather than strict land title (Bromley and Cernea 1989). These leases can be held by an entire community, which then decides how land within the leased area is to be used. Complex legal codes are difficult to uphold, and rigid frameworks may not be sufficiently responsive to local needs. Many local management systems are based on flexible contracts, both implicit and explicit, that may be renegotiated whenever circumstances require. Replacing such systems with rigid laws could be disruptive, and may not result in any net benefits for savanna land management. Generally, legal frameworks should be enabling rather than prescriptive. Firm legal rights to land or natural resources need to focus on key resource areas or definable property, such as trees or water sources; attempts at strict territorial definition will probably fail, as did colonial attempts at boundary delineation.

Incentives for savanna management

Many attempts to encourage group management of savanna areas have failed because insufficient attention was paid to the institutional and legal arrangements necessary to encourage collective management. Establishing social and economic management incentives as the centrepiece of land tenure initiatives is vital, as recognised by the CAMPFIRE programme in Zimbabwe (Murphree and Cumming, this book). Arguments structured simply in terms of "savanna conservation" or "tackling desertification" are unconvincing if economic gains to local people are not anticipated.

Collective management is likely to result in more effective management of high-value land, and it is here that any such initiative might most appropriately start. The development of effective common-property management institutions relies on the existence of common objectives, or at least broadly similar common interests, and a strong dependence on the resource to be managed. Where common interests, resource dependency and production objectives diverge, effective and equitable group management will be much harder.

Planning processes

Participatory approaches to savanna land-use planning need to be encouraged. Local land users should be directly involved in land-use planning and decision-making, as part of a democratic and accountable process. All costs and benefits of different land-use options must be evaluated in the planning process, using a full accounting system that takes explicit account of such factors as environmental and opportunity costs of alternative land-use options. A new form of project assessment that evaluates direct, indirect and non-monetary costs and benefits of different options will have to be developed. The effectiveness of land tenure systems depends on local people having a stake in the planning process (Lane and Scoones, this book).

State and local institutions

There has been much discussion about devolving control of resource management to local communities in the last decade. The argument for local level management (Toulmin 1990) begs a number of questions, however:

(1) At what level should a community be defined? Must new structures be established or will "traditional" institutions suffice? Given the differentiated nature of rural communities, do representative bodies with the capacity for guaranteeing fair access to land exist?

(2)   If control is vested with local villages or herder groups, what support will they require from local government, in terms of legal enforcement of rights and technical support for land management?

(3)   How can conflicts between strategic national interests, such as desire for wildlife revenue or commercial beef ranches, and local priorities be resolved? How should rights be allocated between traditional land users and those who have recently moved into an area?

Some authors have argued for co-management agreements between State and local institutions (Lawry 1990). Under such a system, the State is vested with certain roles such as protecting legal rights, providing technical support and coordinating participating groups, and has responsibility for convening dispute negotiations. Local groups have rights to manage and control local resources. For this system to work, however, the State legislature must be democratically accountable and local institutions must be responsive to local concerns. In many cases, these conditions do not apply.

REFERENCES

Abel, N. and Blaikie, P. (1986) Elephants, people, parks and development: The case of the Luangwa valley, Zambia. *Environmental Management* 10:735–51.

Anderson, D. (1987) Cultivating pastoralists: Ecology and economy among the Il Chamus of Baringo, 1840–1980. In: D. Johnson, and D. Anderson (eds). *The ecology of survival: case studies from northeast African history.* Westview, Boulder, Colorado.

Arhem, K. (1985) *Pastoral man in the Garden of Eden: The Maasai of the Ngorongoro Conservation Area, Tanzania.* Scandinavian Institute of African Studies, Uppsala.

Behnke, R. (1985) *Rangeland development and the improvement of livestock production.* S. Darfur. WSP, Nyala, Sudan.

Behnke, R. (1988) Range enclosure in central Somalia. Overseas Development Institute Pastoral Development Network Paper 25b. Overseas Development Institute, London.

Bromley, D. and Cernea, M. (1989) The management of common property natural resources: Some conceptual and operational fallacies. World Bank Discussion Paper No 57.

Ciriacy-Wantrup, S. and Bishop, R. (1975) Common property as a concept in natural resources policy. *Natural Resources Journal* 15:713–27.

Cline-Cole, R. (1988) Sowing seeds of discord: Induced wet rice cultivation and land-use conflicts in the Hadeija valley, Nigeria. Paper to the ASA conference, Cambridge.

Collett, D. (1987) Pastoralists and wildlife: Image and reality in Kenya Maasailand. In: D. Anderson and R. Grove (eds) *Conservation in Africa.* Cambridge University Press, Cambridge.

Cousins, B. (1988) Community, class and grazing management in Zimbabwe's communal lands. In: B. Cousins (ed) *People, land and livestock.* Proceedings of a workshop. CASS/GTZ, Zimbabwe.

Elbow, K. and Rochegude, A. (1990) A layperson's guide to the forest codes of Mali, Niger and Senegal. Land Tenure Center Paper 139, Madison, Wisconsin.

Grandin, B. (1988) Kajiado Maasailand: the socio-historical context and group ranches. In: S. Bekure, P. De Leeuw and B. Grandin (eds) *Maasai herding: an investigation of pastoral production in group ranches in Kenya.* International Livestock Centre for Africa, Nairobi.

Hardin, G. (1968) The tragedy of the commons. *Science* 162:1243–48.

Helland, J. (1982) Social organisation and water control among the Borana. *Development and Change* 13:239–58.

Hogg, R. (1987) Development in Kenya: Drought, desertification and food scarcity. *African Affairs* 86:47–58.

Horowitz, M. and Salem-Murdock, M. (1991) River basin development policy, women and children: A case study from the Senegal Valley. Symposium on Women and Children First, UN Conference on Environment and Development/UNICEF Preparatory Meeting, April 1991, Geneva.

Jodha, N. (1986) Common property resources and rural poor in dry regions of India. *Economic and Political Weekly* 21:1169–81.

Jodha, N. (1991) Rural common property resources: A growing crisis. *Gatekeeper Series* SA25, International Institute for Environment and Development, London.

Lane, C. (1990) Barabaig natural resource management: Sustainable land use under threat of destruction. Discussion Paper 12, UNRISD, Geneva.

Lane, C. and Pretty, J. (1990) Displaced pastoralists and transferred wheat technology in Tanzania. *Gatekeeper Series* SA20, International Institute for Environment and Development, London.

Lawry, S. (1989) Tenure policy toward common property natural resources. Land Tenure Center Paper 134. Madison, Wisconsin.

Lawry, S. (1990) Tenure policy and natural resource management in Sahelian West Africa. Land Tenure Center Paper 130. Madison, Wisconsin.

Le Bris, E., Le Roy, E. and Mathieu, P. (1991) *L'appropriation de la terre en Afrique noire: Manual d'analyse, de décision et de gestion foncières*. Karthala, Paris.

Little, P. (1985) Absentee herd owners and part time pastoralists: The political economy of resource use in northern Kenya. *Human Ecology* 13:131–51.

Little, P. (1987) Land use conflicts in the agricultural-pastoral borderlands: The case of Kenya. In: P. Little and M. Horowitz (eds) *Lands at risk in the third world: local-level perspectives*. Westview, Boulder, Colorado.

Manger, L. (1987) Traders, farmers and pastoralists: Economic adaptions and environmental problems in the southern Nuba mountains of the Sudan. In: D. Johnson and D. Anderson (eds) *The ecology of survival: Case studies from north east African history*. Westview, Boulder, Colorado.

Moorehead, R. (1989) Changes taking place in common property resource management in the inland Niger delta of Mali. In: F. Berkes (ed) *Common property resources: Ecology and community based sustainable development*. Belhaven Press, London.

Ndagala, D. (1982) "Operation Imparnati": The sedentarization of the pastoral Maasai. *Nomadic Peoples*, No. 10. Commission on Nomadic Peoples, International Union of Anthropological and Ethnological Sciences, Montreal.

Nelson, R. (1990) Drylands management: The desertification problem. World Bank Technical Paper 116, Washington DC.

Oxby, C. (1981) *Group Ranches in Africa*. FAO, Rome.

Parkipuny, M. (1991) Pastoralism, conservation and development in the Greater Serengeti Region. Dryland Network Programme Issues Paper 26. International Institute for Environment and Development, London.

Runge, C.F. (1981) Common property externalities: Isolation, assurance, and resource depletion in a traditional grazing context. *American Journal of Agricultural Economics* 63 (4): 595–606.

Runge, C. (1986) Common property and collective action in economic development. *World Development* 14:623–35.

Sandford, S. (1990) Keeping an eye on the TGLP. National Institute of Development and Cultural Research Working Paper 31, Gaborone.

Schoonmaker-Freudenberger, K. (1991) Mbegué: The disingenuous destruction of a Sahelian forest. Drylands Programme Issues Paper, 29, IIED, London.

Scoones, I. and Cousins, B. (1991) Key resources for agriculture and grazing: the struggle for control over dambo resources in Zimbabwe. In: I. Scoones (ed) *Wetlands in drylands: The agroecology of savanna systems in Africa*. International Institute for Environment and Development Drylands Programme, London.

Shepherd, G. (1988) The reality of the commons: Answering Hardin from Somalia. Social Forestry Network Paper 6d. Overseas Development Institute, London.

Sorbo, G. (1985) *Tenants and nomads in eastern Sudan: A study of economic adaptation on the New Halfa scheme.* Scandinavian Institute of African Studies, Uppsala.

Swallow, B. (1990) Strategies and tenure in African livestock development. Land Tenure Center Paper 140. Madison, Wisconsin.

Sweet, R. (1987) The communal grazing cell experience in Botswana. Pastoral Development Network Paper 23b. Overseas Development Institute, London.

Swift, J., Toulmin, C. and Chatting, S. (1989) Providing services for nomadic people. UNICEF Working Paper, New York.

Thébaud. B. (1990) Politiques d'hydraulique pastorale et gestion de l'espace au Sahel. *Cahiers des Sciences Humaines* 26:1–2.

Toulmin, C. (1990) Natural resource management in the Sahel: Bridging the gap between bottom up and top down. Paper presented to workshop on Human Life in African Arid Lands, Uppsala, September 6–9,1990.

Toulmin, C. (1983) Herders and farmer? or farmer-herders and herder-farmers? Pastoral Development Network Paper 15d. Overseas Development Institute, London.

Touré, O. (1990) Where herders don't herd anymore: Experience from the Ferlo, northern Senegal. Drylands Network Program Issues Paper 22. International Institute for Environment and Development, London.

UNEP (1984) General assessment of progress in the implementation of the plan of action to combat desertification 1978–1984. United Nations Environment Program, Nairobi.

Warren, A. and Agnew, C. (1988) An assessment of desertification and land degradation in arid and semi-arid areas. Drylands Network Program Issues Paper 2. International Institute for Environment and Development, London.

Western, D. (1982) The environment and ecology of pastoralists in arid savannas. *Development and Change* 13 (2):183–211.

Zimbabwe Government (1986) Communal area management programme for indigenous resources (CAMPFIRE). Department of National Parks and Wildlife Management, Harare.

CHAPTER 4

LAND TENURE IN PRIVATE AND MIXED-PROPERTY REGIMES

J.H. Holmes

SUMMARY

In the world's tropical savannas, where nomadic pastoralism is no longer practised, the diversity of land-tenure systems is a logical response to the diversity of land use, stages of economic development, population pressure, social structure and distribution of political power.

In Latin America and India, extreme inequality in land distribution is the dominant issue. Large absentee-owned estates are common throughout many Latin American savannas. India is characterised by smaller scale, but similarly exploitative petty landlordism. Both regions exhibit marked inequities and inefficiencies in resource use. While land reform has been a long-standing policy, it no longer translates into programmes capable of implementation. The experience of the last four decades confirms that entrenched power relations within these societies will prevent effective land reform, either through democratic processes or through revolutionary social upheaval. A major impediment is the lack of cohesive political will among the rural dispossessed, and their increasing irrelevance to the processes of economic and political change.

There are comparable inequalities, based on racial distinctions, in South Africa and, to a lesser extent, Zimbabwe, with land redistribution being a central issue in any shift in power between racial groups.

INTRODUCTION

Access to land resources is a pivotal thread in the fabric of society. The basic structure of any society, save perhaps the most urbanised, can be adduced from its land-tenure regimes and the resultant mode of sharing in

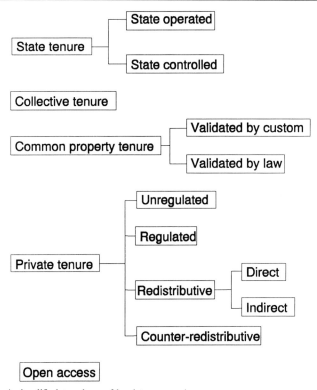

Figure 4.1   A simplified typology of land tenure regimes

landed property. Recently, there has been a growing recognition of the critical role of land tenures, and of associated regimes of property (or resource) rights as the crucial element in pursuit of "optimal" systems of resource use.

Perhaps to a greater extent than any other extensive global biome, the savanna regions show a great diversity in property-rights regimes, as expressed in land-tenure systems. This is indicative of diversity in political, social and economic systems, paralleled by diversity in human settlement, production modes and levels of intensification. Furthermore, the current flux in savanna land use is matched by volatility, conflict and change in land tenure and ownership.

DIVERSITY IN LAND-TENURE REGIMES

Figure 4.1 provides a simplified typology of land-tenure regimes. All of the regimes shown may be found in some form within the savanna regions.

This chapter focuses on areas where, at least, a significant proportion of the land is administered through a private-property regime.

Latin American savannas are dominantly held in unregulated private tenures, mainly as large landed estates. Although nominally redistributive, government policies have in practice been strongly counter-redistributive, as shown later. In the African savannas, the earliest displacements of the common-property regimes were in the two main countries of European settlement, South Africa and the former Southern Rhodesia, where counter-redistributive policies of land privatisation have been pursued.

India's savannas are largely held under private tenures, which were highly counter-redistributive under English colonial rule. Post-independence land reform programmes have been effective only in eliminating the highest levels in the hierarchy of landlords. Petty landlordism remains widespread, and is now being reduced, more through the negative effects of population pressure than through positive efforts at land reform. While all villages in the Indian savannas have traditionally held significant tracts of common lands, these have been persistently eroded by privatisation, State intervention and management collapse (Gadgil, this book).

Australia's savannas have been held predominantly under pastoral leasehold tenure, which is a *de jure* State-controlled and State-owned tenure, but a *de facto* regulated private tenure. Since 1976, Australian land policies have taken a radically new direction with the award of Aboriginal non-transferable freehold tenure, which generally appears to be emerging as a *de facto* common-property tenure for traditional uses and a collective tenure for non-traditional uses (Holmes and Mott, this book).

CONFLICT AND CHANGE IN LAND USE AND LAND TENURE

The tropical savannas are also noteworthy for their recent instability and change in land tenures, often arising from conflict between irreconcilable interest groups, each identified with a distinctive mode of land ownership and use. Commonly, the conflict is linked to land-use intensification and land privatisation involving the displacement of more extensive modes of occupancy and traditional tenures.

LATIN AMERICA

Apart from limited areas of higher productivity and greater accessibility in the Brazilian *cerrado*, which are subject to the twin pressures of surplus rural population and capitalist penetration, the savannas of Brazil, Paraguay and Venezuela still preserve many of the more traditional elements of Latin

American land occupancy, land tenure and land ownership. These elements include: extreme concentration of land ownership and wealth; the use of land monopoly to ensure labour capture, accompanied by limited labour demand and institutionalised labour surplus sustained by the dominantly subsistence activities of the *minifundistas*; emphasis on land as a source of wealth and power; extensive land use in the traditional *hacienda* mode with retarded capitalist penetration; some poverty and destitution among *minifundistas* and landless workers and, in Brazil, inclusion of a "frontier" zone of colonisation and conflict (de Janvry 1981; Grindle 1986).

The State's counter-redistributive role

At least outwardly, at some stage over the last 30 years, almost every Latin American government has legislated for agrarian reform, using direct redistributive mechanisms of land purchase and resettlement. As shown below, these policies have had negligible impact. Token efforts at redistribution continue to be outweighed by counter-redistributive mechanisms, either *de jure* or *de facto*, which remain entrenched in rural Latin America.

There is a long-standing tradition of awarding to large landholders an array of local economic, civic, police and judicial powers within and around their own domains. While the *de jure* award of such powers may well have been markedly reduced, they still remain entrenched in daily affairs, accepted by petty officials and by the ill-educated, impoverished rural population at large, even if grudgingly.

The principle of the sanctity of private property has to be treated with caution. There is adequate legislation with regard to land-title identification and registration, but it is not enforced. Hence the "climate of insecurity" is maintained in land as well as in labour relations. Seemingly perversely, and greatly to the advantage of the landholders' dominance of local power, is the entrenched belief of the underprivileged peasants in the sanctity of private-property rights, respect for which is considered to be a citizen's most important duty (Lindqvist 1979).

Taxation or rating schemes to encourage the redistribution of land to small-scale investors are uncommon. Land taxes and death duties are rarely found in Latin American legislation. In 1961 as the result of an amendment to the Brazilian federal constitution, the power of land taxation was passed from the states to the *municipios* as a means of financing local municipal works. As stated by the Pan American Union (1966) this has provided one further incentive to maintain inadequate local land registers and leave local landlords with responsibility for defining "existing" rights. The collection of taxes and rates on rural land continues to be negligible, with the burden being transferred to urban land and businesses, and with *municipios* being unable to provide adequate services (Webb 1974).

Regressive effects of agricultural support schemes

Up to the 1960s, many Latin American states, most notably Brazil, pursued policies of import substitution, focussing on accelerated industrialisation, to the neglect of agriculture, particularly of the internal food-production sector. Continuing high import bills, with ensuing crises in balance-of-payments, led to a reorientation towards export promotion, including an emphasis on large-scale agricultural production, from 1966 onwards in Brazil. The extraordinary array of anomalies in these programmes and their inefficient and highly inequitable outcomes are attributed by Grindle (1986) to the actions of the State. She observes that "governments regaled commercial agriculture with policy incentives showing little regard to the negative repercussions on the lagging part of agriculture which produced foodstuffs".

Subsidised loans for mechanisation, fertiliser use, fuel and other inputs, often at interest rates far below levels of inflation, encouraged large landholders to engage in highly irrational and inefficient forms of labour replacement (see also Klink et al., and Silva and Moreno, this book). Assistance from the World Bank and other international agencies was channelled into wasteful expenditure on large estates or often diverted into urban or speculative investments (Sorj 1980) while over 80 per cent of landholders remained ineligible for any financial assistance. Similar imbalances existed in research and extension services. Even relief programmes, intended to ameliorate the outcomes of drought and rural impoverishment were redirected to favour the wealthy, being channelled into irrigation schemes and low-cost labour employment on "work fronts". Lassen concluded that "perhaps nowhere in the world are large farmers as coddled as they are in Brazil by generous government subsidies in the form of tax incentives, extremely soft loans and price supports" (Lassen 1980, quoted by Grindle). Noting the adverse social, economic and environmental impacts of such programmes, both of the Latin American chapters in this book recommend a transition to a more market-based approach which reveals the nature of the ecological and economic constraints on such development.

The failure of agrarian reform

Until 1960, only three Latin American nations had attempted agrarian reform, and in each case it was linked to violent social revolution. However, between 1960 and 1964, 14 States enacted reform legislation, including the savanna States of Venezuela (1960), Paraguay (1963) and Brazil (1964), as pointed out by Grindle (1986).

In all cases, seemingly substantial, ambitious reform programmes were legislated, administrative bureaucracies established, funding provided and some land purchased for redistribution. In implementation, however, not only were all these programmes wasteful, misdirected and generally abused but also they were rapidly reduced to a state of tokenism and near-paralysis. The rapid demise of land reform can be attributed to various factors, including the high costs associated with overgenerous purchase prices for large holdings, faulty administration and abuse, ineffective project development, and the rapid replacement of reformist-minded governments. As indicated by the case studies in this book, in each case the programmes were doomed to failure simply because they were not accompanied by the macro-economic and institutional reforms necessary to prevent the socially inefficient and inequitable forms of land distribution re-emerging. As Silva and Moreno (this book) note dramatically, in Venezuela agrarian reform ultimately resulted in more, not less, land under the control of large estate owners.

Appropriation of the frontier

In Brazil, population pressures, aggravated by extreme maldistribution of land, are leading to spill-over effects, on a much broader scale, in the nation's rainforest and savanna regions. Token efforts at agrarian reform, initiated in 1964, were soon displaced by government-sponsored colonis-ation programmes for the Amazon rainforests and adjacent savannas, which were regarded as a convenient alternative outlet for the ever-expanding legions of rural landless workers. However, even in the frontier regions, both the indigenous peoples and the colonising *minifundistas* have been pushed aside by aggressive, often violent campaigns of land capture by the more powerful. It is mainly in these frontier regions, of great flux in land use, land tenure and land ownership, that the latent conflict, endemic within Brazil's rural structure, is actively engaged. Early government goals of planned colonisation by previously landless smallholders have been swept aside, with the policy change being rationalised in terms of the need to attract private capital to the frontier (Klink et al, this book).

INDIA

The Indian tropical savanna zone coincides almost exactly with the Indian peninsula, save for the rainforested Malabar Coast and adjacent southern uplands. The original savanna vegetation has almost entirely disappeared under the pressures of prolonged, intense human occupation. Human inputs into production systems ensure a relatively high output from cropping systems, sustained by labour-intensive *landesque* capital, invested in wells, tanks, ponds, channels, drains, padi and dryland terraces, forming a human-altered landscape divided into an intricate pattern of small landholdings.

In contrast to Latin America, current Indian landlordism is mainly a small-scale, village-level phenomenon, with a numerous, politically powerful class of petty landlords of modest means able to maintain an exploitative, *rentier* form of land tenure through their political influence within local and State party machines, aided by the lack of political or economic power within the large rural underclass of tenants and labourers.

Within this rural landscape dominated by private holdings, almost all villages still retain some common land serving a multiplicity of purposes. Commons are traditionally attached to individual villages, providing community pastures, community forests, waste lands, common dumping and threshing grounds, watershed drainages, ponds, river banks and beds. While accurate data on land tenure are not available, both Agarwal and Narain (1989) and Jodha (1990) observe that there has been a marked reduction in common lands since 1951, but that the savannas still retain a higher proportion than do the humid plains.

Pressures on the commons

Given these unequal power relations, reinforced by land hunger, it is hardly surprising that the village commons have experienced "area shrinkage, productivity decline and management collapse" (Jodha 1990).

Generally, and as observed by Gadgil (this book), the State has had a negative role in promoting sustainable forms of savanna land use. It has done this by whittling away the common lands and by destroying community management. As in Venezuela, "populist programmes" of land reform were redirected towards the commons, which proved easy targets but, though undertaken in the name of the poor, these state-sponsored programmes of privatisation mainly benefited other landholders (Jodha 1990). The second form of destructive State intervention was through programmes for raising the productivity of the commons, often focussing on forestry and emphasising production technology rather than community involvement. State efforts at streamlining local authority structures also eroded community institutions.

... traditional conventions and informal social sanctions that regulated common property use were replaced by unenforceable legal and administrative measures. This has marginalised the people's initiatives and alienated them from the resource base. At best, the commons are mismanaged and underproductive, at worst they become open access "wastelands" providing sparse pickings for the destitute (Jodha 1990).

As a result, Gadgil (this book) recommends the removal of government enterprise from the commons, and the devolution of responsibility for them back to local communities.

> The evidence in the country clearly shows that where people are given control and a vested interest is created in the development of the natural resource base, people have worked together to bring about a rationality in the use of natural resources (Agarwal and Narain 1989).

AFRICA

Spurred by local, national and international imperatives the African savannas are currently experiencing great instability and change in land tenures. The most potent changes are linked to the transformation of land into a marketable commodity, and intervention by emergent nation-states, with an ideological commitment to national economic growth and social change. Three distinctive contexts leading to land-tenure conflict and change can be observed: traditional sources of conflict and change; those arising from European settlement and land acquisition; and current ones caused by the disruption of traditional systems, involving commoditisation of land and labour, often actively supported by the State.

Traditional conflicts and complementarities in tenures

While common-property regimes appear to have been near-universal in African savannas, there was considerable variability in resource allocation methods within these regimes with varying degrees of attachment of individual resource rights. Partial "privatisation" was associated with more sedentary systems of permanent cultivation.

Scoones, Toulmin and Lane (this book) discuss various forms of conflict over land tenure and use. Although conflict did occur between groups who engaged in different modes of land use, most notably between pastoralists and cultivators, it should be recognised that more persistent latent and actual conflict occurred between rival groups engaged in the same modes of resource use.

In this book, the chapters by Lane and Scoones, Kituyi, Murphree and Cumming and Pearce provide insights into traditional common-property

74

regimes, and the confused mixture of property regimes that co-exist in any one region. It is now clear that complex linkages exist between resource allocation, management, customary rights, obligations and sanctions and religious beliefs, either were not understood or, intentionally, ignored. With hindsight, and as indicated in the preceding chapter, it is now clear that many common-property regimes are more equitable, more productive and less likely to cause land degradation than private-property regimes. The challenge now is to find equitable ways to undo past mistakes and, at the same time, provide a foundation which will enable local development.

Conflicts arising from European land acquisition

European settlement within Africa's savannas was almost entirely confined to the subtropical uplands of the south, encompassing South Africa and Zimbabwe, with a near-monopoly of the more productive lands. The rigidly apartheid structure of South African society has its underpinnings in a racially restrictive system of land ownership, with 87 per cent of all land being reserved for Whites, who currently comprise only 13.4 per cent of the population. The outcome has been an extreme maldistribution of resources, comparable to that in Latin America and with broadly similar outcomes. As observed by Mentis and Seijas (this book), the great mass of the rural population comprises a powerless underclass of unskilled, undereducated labourers, subordinated within a coercive system of labour capture. The majority of the Black underclass are restricted to designated homelands where a communal subsistence peasant economy, sustained by women, children and the elderly, acts as a support system which further assists labour exploitation by the dominant White class in a manner broadly comparable to *minifundista* farming in Latin America.

A similar, if slightly less unequal, system of land acquisition and maldistribution also occurred in Zimbabwe. In both countries, as in Latin America, inequalities were magnified by government assistance to White agriculture in the form of subsidies, drought assistance, research, extension and cheap loans.

Compared with Latin America, the South African system of unequal relations has been more institutionalised within a system of laws founded upon a sharp racial demarcation, with tight restrictions on residency and work permits, by which a mainly male workforce can be harnessed to designated jobs in fixed locations, and returned to homelands when not required. Although overcrowded and under-resourced, the homelands provide an important resource, not only for subsistence, but also for cultural survival. According to Mentis and Seijas (this book), the homelands remain reasonably managed on a sustainable basis and with established property-rights systems, comprising a mix of private and common regimes which continue to be managed effectively in spite of population pressures and the dislocations imposed by labour capture.

The process of dismantling the legislated structure of apartheid was taken a further stage in March 1991, with the government's declaration that it would abolish the Land Acts, thereby removing racial restrictions on land-ownership. However, the existing extreme maldistribution of ownership and wealth, together with *de facto* apartheid, is likely to be maintained, thereby shifting South Africa closer to the Latin American situation unless an active redistributive policy is adopted. Whatever form of government emerges in the near future, the question of agrarian reform will certainly be one of the leading and most contentious issues. Noting the failure of past agrarian reform programmes, Mentis and Seijas argue for a market-based transition to a new regime, possibly aided by a land tax to increase the pace at which wealth is redistributed. They envisage a new economy where blacks acquire wealth and then land. They do not address the question of whether or not part or all of the homelands should be privatised. As Kituyi (this book) notes, common-property regimes make it more difficult for individuals, as distinct from the community, to acquire wealth and use the means to finance further investment.

Murphree and Cumming provide an informed appraisal of the actively redistributive policies of the Zimbabwean government which, from 1980 to 1990, adopted a gradualist approach, with an emphasis on State or collective tenure systems. This has now been replaced by an accelerated programme using both direct and indirect redistributive measures. While the authors are supportive of redistributive policies, they express disquiet at the undue focus on the issue of State control versus private ownership, with inadequate attention to the revival of common-property tenures. They also are concerned about deficiencies in programme implementation. In Botswana there has been a gradual process of change in resource rights and use on common-property pastures, triggered mainly by local privatisation of water and of land (Pearce, this book; Peters 1987). The sinking of bore-holes, initially government-funded, at first led to expansion of the range, greater herd mobility and a phase of virtual open access. This has been gradually replaced by private ownership of bores and the treatment of water as a commodity. As mixed private-communal grazing regimes emerge and herd mobility is reduced, the payment of watering fees and the purchase of shares in syndicates that "own" watering points are becoming more common. Although this partial privatisation leads to progressive loss of the communal system, nevertheless it is not inconsistent with long-held notions of the Tswana: that anyone putting effort or investment into land is due the benefit.

AUSTRALIA

Of the four major savanna zones described in this chapter, the two with the closest parallels in modes of human occupation and land tenure are Australia and Latin America. There are strong similarities in the process of initial colonisation and in the ensuing prolonged phase of retarded development based on extensive cattle-raising. Australia's tropical savannas were occupied by European ranchers during the late nineteenth and early twentieth century, being the final frontier in the colonisation of Australia's grazing lands, a process that closely paralleled Latin America's earlier wave of extensive land settlement. Occupation was aided by the doctrine of *terra nullius* as a basis for dispossession of the scattered, ill-defended indigenous peoples. Vast tracts of land were occupied under government-issued leasehold title, providing the foundation for the cattle stations of northern Australia. These stations were (and still are) characterised by large land areas under extensive occupancy, with low capital investment, low labour demand and little capacity to promote regional economic development.

One further marked similarity was in the coercive form of labour capture through territorial dispossession, by which Australia's northern cattle stations acquired an Aboriginal workforce of stockmen, labourers and domestic servants, almost exactly replicating the Latin American and, to some extent, the South African *modus operandi*. Insulated from the Australian mainstream through remoteness and an underdeveloped economy, the more remote northern livestock regions also remained socially backward.

In recent decades, however, North Australian land allocation policies have been in marked contrast to those in Latin America. These are the outcome of belated regional application of Australia's strongly egalitarian philosophy of land ownership which originated from closer settlement processes during the late nineteenth century and was facilitated under the pastoral leasehold systems. Core elements were:

(1) Restricted property rights, specifically rights to make use of pastures and water for pastoral purposes only, together with basic rights to occupy and enjoy the land;

(2) Specified obligations, which in due course could include obligations to reside on the land, to undertake investments and to maintain minimum numbers of livestock; and

(3) Extinguishable rights, with leases being for fixed terms, the State retaining the right of non-renewal on expiry with compensation only for fixed capital investments.

Change was heralded in 1967 with the award of full citizenship rights to Aboriginals, and, since then, change has been remarkably rapid as reformist national goals, espoused by metropolitan Australia, have been imposed upon the remote, undeveloped regions.

As in Canada, the land rights legislation of the 1970s and 1980s has introduced entirely new concepts of land ownership. The most radical change has been the introduction of non-marketable group-owned freehold titles. Another major innovation has been in the award of an expanded set of property rights, often including full title to subsurface resources such as minerals, together with powers to restrict access, greater than those held under normal freehold title.

A form of collective property tenure is emerging in the sharing of resource rights (and revenues) from non-traditional uses among all members of the ownership group, with membership being based upon group acceptance. This is distinctly different from the form of corporate ownership applied by the United States government in its Alaska Native Claims Settlement Act 1971 which established corporations within which individual shares were allocated, with ongoing debate concerning whether these shares should be marketable and, if so, whether only to other corporation members or to outsiders.

Also incorporated into Australian land rights legislation is the recognition of traditional property rights relating to religious, ceremonial and subsistence activities. There are close parallels with common-property systems within which clearly specified hereditary individual resource rights are imbedded within a group's shared territory.

As pointed out by Holmes and Mott (this book), the long-term outcomes from the award of Aboriginal land rights still remain unclear. The possibility of modified tenures outside of Aboriginal "homelands" enabling some mix of Aboriginal and non-Aboriginal uses, especially within national parks and other reserves, is most promising.

CONCLUSION

Given the similarity of the natural environment in the four major tropical zones discussed in this chapter, the marked divergence in land-tenure systems is a striking indicator of the sharp differences in the direction and pace of socio-economic development and in population pressures. Almost the entire spectrum of land-tenure regimes (see Figure 4.1) is represented in a substantial way in at least one of the savanna zones.

Amid these highly differentiated land-tenure regimes, there has been a common thread of recent instability and conflict, demanding policies directed towards land redistribution and/or land-tenure change. In all cases there has been significant legislative action, giving the appearance of a commitment to reform. However, wherever direct programmes of land redistribution were required and enacted, the outcomes have been universally modest, and the momentum to reform has soon been dissipated. Whether in Brazil, Venezuela, South Africa or India, commentators have all agreed that substantial land reform will only be achieved through a marked shift in power relations, requiring revolutionary political and social change. Even with radical change, land reform may still not be accomplished, given the increasing marginalisation of the rural dispossessed.

The ideal system of land tenure is that which provides a harmonious matching between resource use and resource rights, whether the mode of resource use is based upon private, collective, common-property or some other organisational mode.

ACKNOWLEDGEMENTS

This chapter was conceived during the Nairobi workshop, in January 1991, and benefited from the discussions at that workshop. My thanks are due to the workshop participants and especially to Michael Young, Marshall Murphree, Charles Lane and Ian Scoones.

REFERENCES

Agarwal, A. and Narain, S. (1989) *Towards green villages: A strategy for environmentally-sound and participatory rural development*. Centre for Science and Environment, New Delhi.

de Janvry, A. (1981) *The agrarian question and reformism in Latin America*. Johns Hopkins Press, Baltimore.

Grindle, M.S. (1986) *State and countryside: development policy and agrarian politics in Latin America*. Johns Hopkins University Press, Baltimore.

Jodha, N.S. (1990) Rural common-property resources: Contributions and crisis. *Economic and Political Weekly* June 30, 65–78.

Lindqvist, S. (1979) *Land and Power in South America*. Penguin, Harmondsworth.

Pan American Union (1966) *Land tenure conditions and socio-economic development of the agricultural sector: Brazil*. Washington DC.

Peters, P.E. (1987) Embedded systems and rooted models: The grazing lands of Botswana and the commons debate. In: B.J. McCay and J.M. Acheson (eds) *The Question of the Commons: The Culture and Ecology of Communal Resources*. University of Arizona Press, Tucson.

Sorj, B. (1980) Agrarian structure and politics in present day Brazil. *Latin American Perspectives* 7 (1):23–34.

Webb, K.E. (1974) *The changing face of northeast Brazil*. Columbia United Press, New York.

CHAPTER 5

NATIONAL AND INTERNATIONAL INFLUENCES THAT DRIVE SAVANNA LAND USE

M.D. Young

SUMMARY

National and international policies are often formulated with little attention to their implications for savanna land use. As a result, many of them tend to increase rates of land degradation, encourage inefficient forms of development and cause social conflict.

Examples include exchange rate controls that suppress export opportunities and macro-economic policies which, by increasing the real interest rate, make resource conservation a suboptimal strategy. International trading policies, especially preferential trading agreements, tend to reward overgrazing and have inequitable consequences for many savanna people.

At the national level, preferential tax concessions and input subsidies tend to diminish the likelihood of maintaining the environmental integrity of savanna ecosystems, ensuring socially efficient forms of resource use and making the distribution of resource rights more equitable. In most cases, the interests of savanna land users and the savanna itself are best served by reliance upon mechanisms that couple market forces to the natural economic and ecological capacity of savanna ecosystems.

INTRODUCTION

The world's savannas are well known for their colourful people, attractive wildlife, open landscapes and warm climates. Many of them are also well known for the extensive land degradation that has occurred and still is occurring.

Rainforest management was the environmental issue of the 1980s and global change is likely to be the issue of the 1990s. At an international level, the health of the world's savannas is no longer a focal point for discussion. Other issues like global change and rainforest management now dominate international policy debates. But savanna management – flying under the banner of desertification – is an issue of the 1970s that has not gone away. The immense problems of this part of our global ecosystem were talked about but not solved. The causes of most of the problems observed in the 1970s remain.

Major land uses within the savannas include pastoralism, fuel-wood collection, some cropping and a few specialist industries such as the collection of gum arabic. Whilst the condition of the world's savannas may seem irrelevant to those who participate in international decision-making processes, many international and national decisions appear to dominate the land-use strategies adopted by savanna people. Note, however, that most of the policy decisions that affect savanna land use are made for reasons that have little, if anything, to do with it. The pathways of influence of international conditions and policy decisions, however, are complex and poorly understood.

This chapter focuses upon some of the international causes of resource degradation and national opportunities to improve the welfare of savanna people and the land they use. Recognising the dangers of being overly simplistic, it attempts to unravel the effects of national and international policy on savanna land use. The aim is to search for summary principles and recommendations that, if adopted, would improve the welfare of savanna people and the resources that support them. Fortunately, there is nothing particularly special about savanna resources. If implemented, the policy recommendations that follow would improve investment and resource use throughout much of the world.

KEY DRIVING FORCES

At the national and international level, the key factors that influence savanna land use, indeed drive many decisions, include:

(1) International trade policy, particularly tariff and import quota arrangements;

(2) Exchange-rate policies that change relative prices and costs;

(3) International aid policies;

(4) The wide array of price support, production subsidy, input subsidy and other arrangements that influence the demand for savanna products; and

(5) The existence of a host of arrangements that tend to underprice the real social value of biomass, biodiversity and environmental quality.

INTERNATIONAL TRADING ARRANGEMENTS

International trading arrangements affect the prices received for savanna products and, through this influence, the nature of savanna land use. Generally, higher prices increase development options but, when subsidies and other mechanisms are used to maintain high prices, market information about ecological feedback on resource capability is lost. Undistorted trading arrangements enable regions to specialise in the production activities most suited to their resource endowment and, hence, gain greater benefits than would otherwise be possible. In economic jargon this is known as comparative advantage. Selling beef to the USA and importing education and medicine is one example of the benefits that savanna people can derive from trading with people who live in other parts of the world. But comparative advantage does not necessarily equate with competitive advantage. World trade is driven by actual prices which, because of market and government failure, often do not reflect the full value of environmental quality and the adverse impacts of land degradation on other people.

Probably the greatest adverse effect of trade-related policies, however, is the aggregate impact of export subsidies, tariff barriers and production subsidies in other countries on the prices savanna people receive for their products. Generally, savannas tend to have a comparative advantage in livestock production. But global markets for these products and their crop substitutes – corn, wheat and sorghum – are highly distorted (see Table 5.1). In the European Community, tariffs (e.g. import quotas and production subsidies) keep beef and grain prices well above world-market prices, with the consequence that the world prices for savanna products are much lower than they otherwise would be. Note also that the tendency of such countries to dump surplus grain and meat on world markets further depresses the income received by savanna people. As a result savanna land use in countries like Australia, Argentina, Brazil, Mali, Tanzania, Venezuela and Zimbabwe is marginalised. Thus, irrespective of domestic policy initiatives, until international arrangements improve, or are compensated for, savanna people must be expected to make inappropriate investments and not use these landscapes in the most sustainable manner possible.

The general and widely accepted conclusion is that freer agricultural trading arrangements (i.e. higher world prices) could be of direct benefit for savanna land use. The use of the word "could" rather than "would" is intentional. As indicated in other chapters in this book, for the anticipated gains to improve the welfare of all, rather than a few, savanna users,

Table 5.1 Average producer subsidy equivalents (PSE) for selected savanna products in various countries 1982–1986. A PSE of + 50% means that for every US$100 that a producer receives, $50 comes via production subsidies, tariff barriers that raise domestic prices etc. A negative PSE means that prices are kept below world-market prices. Source: Economic Research Service (1988), various tables

| Country | Commodity | | | |
	Beef and Veal	Corn	Wheat	Sorghum
Australia	6.4	–	6.8	–
Brazil*	– 33.1	4.0	63.4	–
EC-10*†	44.6	24.8	38.4	–
Japan*	59.0	–	97.8	–
USA	8.7	27.1	36.5	27.1
South Africa	– 50.3	18.3	50.3	–
Nigeria	– 2.8	– 18.7	2.8	–
India	–	–	– 35.3	–

* Beef, not beef and veal
† The European Community excluding Portugal and Spain

changes to property-right regimes plus national taxation and marketing arrangements may also be necessary.

PREFERENTIAL TRADING ARRANGEMENTS

Many savanna countries have signed trading agreements that give them preferential access to protected markets. Perhaps the most studied of these agreements is the Lomé Convention between the European Community and a number of countries in Africa, the Caribbean and the Pacific. Designed to rationalise a variety of trading agreements with colonies of countries that now form the European Community, the Lomé Convention grants participating countries rights to export certain industrial and agricultural products to the European Community at a zero or nominal tariff. Botswana, for example, initially negotiated a 90 per cent tariff reduction on beef exports to the European Community for 6 months and then obtained a series of semi-annual and annual extensions. This trading arrangement has now settled into an annual tariff-free beef quota of 19 000 price-supported tons of beef that may be exported to the European Community (Yaeger 1989) and gives Botswana beef producers access to a market whose prices are 45 per cent above the world price. In effect, it is a direct subsidy or transfer payment from European Community consumers to certain members of the Botswana beef industry. Strict health standards apply and most of the benefits flow to a relatively small number of pastoralists, wholesalers, slaughterhouses and transport operators (Pearce, this book).

Interestingly, all preferential trade agreements tend to increase the price offered to producers in a manner similar to that achievable under free trade. Yet the popular literature is full of criticisms of the Lomé Convention. Yaeger (1989), for example, observes that "the European price subsidy continues to reward ranchers for maintaining large herds of economically marginal cattle on a deteriorating pastoral commons". Assuming that the observation is correct, the question that must be asked is: Why has this preferential trade-induced price increase led to further degradation? The most likely explanations are complex but probably due to other deficiencies in the composite set of conditions that drive investment decisions. In particular, it is likely that:

(1)   The Lomé Convention aggravates the adverse impacts of inappropriate government investments in infrastructure development and tenure arrangements; and

(2)   The price increase is selective and has not been matched by compensating changes in other relative prices and costs[1].

The most general observation that arises from work in this area is that preferential trade agreements, which are limited to one or two commodities, usually have adverse environmental costs that are often greater than the increased revenue achievable through the agreement.

Note also that these agreements tend to be inequitable. Whilst a few privileged resource users gain market access, often the costs imposed on other resource users as a result of such agreements are significant. In the Botswana case, for example, most Lomé Convention benefits go to a small number of wealthy pastoralists whilst the rest have been forced to subsist on a reduced area (Pearce, this book). Moreover, because the resultant resource use is less efficient, all people lose. The question then is: What trading arrangements are likely to be of greatest benefit to the long-term development of the world's savannas? In order of preference the changes likely to be of greatest benefit are:

(1)   A *gradual* movement towards freer trading arrangements;

(2)   The emergence of large trading blocs which allow free trade of goods

---

[1]It is also possible that the price offered is much higher than would be the case under a free trading agreement. It might be a case of overshoot rather than correction.

within them and, because of their size, are able to negotiate favourable arrangements with other trading blocs; and

(3) The development of trading arrangements that exploit a country's comparative advantage without mistaking competitive advantage for comparative advantage (Young 1992).

INTERNATIONAL TRADE RULES

Under the rules that govern international trade there is little which can be done to overcome the tendency of world trading arrangements to reward resource degradation. It is well recognised, however, that trade restrictions and other price support mechanisms force developing countries to either withdraw from global markets or, alternatively, degrade their resources. The result tends to be declining, rather than increasing, standards of living, and hence further incentives for resource degradation. If, however, a mechanism could be found that would enable importing nations to provide an economic incentive for exporting countries to improve resource management standards and increase foreign exchange earnings simul- taneously, then many of the environmental problems caused by international trade could be overcome.

One way of doing this would be to modify the General Agreement on Tariffs and Trade (GATT) so that countries could favour trade with nations that set resource management standards equivalent to their own. Under such an arrangement, all nations that demonstrated compliance with prespecified criteria for sustainable savanna land use would be given preferential access to markets where higher resource management standards are set. In effect, such an arrangement would give them a preferential access to a specialised market niche and consequently an economic incentive to favour sustainable forms of resource use. Note, however, that entry rights to this market would have to be competitive and open to any who set equivalent standards for resource management and environmental protection (Young 1992).

EXCHANGE-RATE POLICY

The linkage between exchange policy and land resource development is similar to the relationship between trade policy and savanna land use. Few countries undervalue their currency but many developing countries use exchange-rate controls to overvalue their currency. Setting official rates above the natural exchange rate depresses export opportunities and, hence, investment in exporting industries. This, in turn, diminishes the derived demand (value) for savanna resources and discourages resource maintenance (Southgate 1989). The reason for this latter problem is that, as land values diminish, fewer and fewer resource-conserving technologies remain profitable. As experience in Brazil demonstrates (Klink et al., this book), when exchange rates are overvalued, ultimately the only remaining strategy is one that, in effect, gradually degrades (mines) the resource base[2].

Besides reducing prices paid for tradable commodities, currency overvaluation lowers the prices paid for imported agricultural inputs like tractors and fertilisers. In addition, because overvaluation favours the substitution of manufactured inputs for labour, local employment opportunities are reduced. Often currency exchange restrictions and import quotas are used to off-set these costs but, when this is done, opportunities for economic growth tend to be reduced but not uniform across all sectors. One can reason, however, that as most savanna conservation practices tend to be labour intensive, currency overvaluation works against savanna development and the welfare of savanna people.

At a totally different level, currency valuation, because it reduces the price of imported farm machinery and makes it attractive to substitute imported fertiliser for livestock manure, promotes the displacement of pastoralists (who purchase few imports) in favour of agricultural development programmes. Note also that the evidence available in this book suggests that, when this occurs, the displaced agricultural labour often moves into more environmentally fragile areas and causes more degradation (Lane and Scoones, and also Kituyi, this book).

Thus, the tentative policy conclusion is that exchange-rate controls have adverse affects on savanna land use and discourage forms of investment most likely to protect and/or increase resource productivity.

NATIONAL MACRO-ECONOMIC, MARKETING AND INPUT SUBSIDY PROGRAMMES

As experience in Brazil illustrates, many of the adverse effects of international policy are aggravated by national macro-economic, taxation and development policies. Taxation concessions, for example, give an economic advantage to those who make profits in other areas and gradually drive local

---

[2]There are occasional exceptions to this general observation. In the savanna areas of West Africa, for example, high exchange rates tend to discourage groundnut production in highly erosive savanna areas. Thus, in special circumstances and in the absence of well-specified resource rights and/or effective land-use regulations, currency overvaluation may prevent savanna degradation.

people out of the land market. Likewise, high real interest rates tend to discourage investment and make resource conservation less profitable. Thus, in Brazil where the inflation rate has averaged 40 per cent per annum and income earned from savanna areas has been exempt from income tax, savanna land has been a highly attractive investment for wealthy people who own businesses which make large profits elsewhere. As a direct result of these macro-economic policies, much savanna land is owned by absentee landlords, and is used and developed inefficiently. There is continual pressure for land reform (Klink et al., this book).

National food and agricultural development policies often have similar adverse impacts on savanna land use and occur in two seemingly opposite directions. In some countries, like Zimbabwe, for example, import restrictions and exchange rate controls coupled with national market restrictions are used to suppress urban food prices (Murphree and Cumming, this book). Inefficiently and inequitably such arrangements tax savanna land users, encourage them to move to cities and, because savanna management options are reduced, increase the likelihood of land degradation.

Taking the opposite approach other countries, in the pursuit of food self-sufficiency and agricultural development, subsidise agriculture inputs and sometimes guarantee agricultural product prices. Whilst politically attractive, such arrangements decouple market signals from the resource base, encourage specialisation beyond ecological capability and therefore tend to increase the likelihood of land degradation.

Input subsidies and price-support arrangements can also drive investment into marginal areas that are unable to support long-term production. In Venezuela, for example, despite 30 years of government subsidies, programmes and substantial infrastructure investment, agriculture is still unable to compete with imports and remains inefficient (Silva and Moreno, this book). The perverse effects of many subsidy programmes also have to be watched with care. For many years, Brazil has paid subsidies in proportion to the area that is cropped with the result that much more land has been cleared than would otherwise be the case (Klink et al., this book). Similarly and as the Barabaig's experience in Tanzania demonstrates, input subsidies can also destroy the viability of highly successful and ecologically sustainable savanna land-use strategies (Lane and Scoones, this book).

CONCLUSION

Unfortunately, many of the policies which have greatest influence, like international trading rules and exchange-rate policy, tend to be formulated with little, if any, consideration of their implications for the world's savannas. This chapter set out to demonstrate how savanna land use is influenced by national and international policies. The main point made throughout is that many of these policies have a vital bearing on the decisions made by savanna people and, with few exceptions, tend to mask ecological feedbacks and thereby increase the likelihood of savanna degradation.

Given all the above, the most general recommendation that can be made is that, as far as possible, governments should avoid subsidising savanna land use and should also be very wary about subsidising crop production in adjacent areas. Global experience suggests that, with few exceptions, most national and international policies that give special preference to one sector or commodity result in inefficient land use, reduce opportunities for sustainable land use, increase land degradation and increase social inequity.

REFERENCES

Economic Research Service (1988) *Estimates of producer subsidy equivalents: Government intervention in agriculture, 1982–86*. United States Department of Agriculture, Washington DC.

Southgate, D. (1899) The economics of land degradation in the third world. Environmental Department Working Paper No.2. World Bank, Washington DC.

Yaeger, R. (1989) Democratic pluralism and ecological crises in Botswana. *The Journal of Developing Areas* 23:385–404.

Young, M.D. (1992) *Sustainable investment and resource use: equity, environmental integrity and economic efficiency*. Parthenon Press, Carnforth, UK and Unesco, Paris.

Section 3

Common-property grazing systems

CHAPTER 6

BARABAIG NATURAL RESOURCE MANAGEMENT

C. Lane and I. Scoones

SUMMARY

For nearly 150 years the Barabaig have managed the Hanang plains in a rational and sustainable manner and environmental damage has been minimal. Key elements of their land management are: protection of the grazing resource from encroachment, conservation of key grazing areas for periods when the potential of other areas has been exhausted, prohibition of permanent residence in some of these areas, and enforcement of these community decisions.

This highly developed communal-grazing system is now being destroyed by official failure to recognise and preserve its benefits. With aid from Canada, the Tanzanian Government is subsidising non-Barabaig people to farm the key grazing areas. Customary rights are becoming unenforceable and defensive action by the Barabaig is prohibited. As a consequence, the sustainability and productivity of the Barabaig system is breaking down. Unknowingly, government policies and programmes are turning parts of the Hanang plains into an unsustainable open-access system: creating the "tragedy of the commons".

If the Barabaig are to survive as a people then new property-right systems that build upon the principles that underpin their traditional grazing system, and that are capable of evolving over time, need to be developed.

INTRODUCTION

In recent years, attention has been given to the form of pastoralists' management strategies. An increased understanding of common-property management systems in extensive rangeland areas has emerged (National Research Council 1986; Lane and Swift 1989; Baxter 1989). African pastoral production has been shown to provide good returns relative to other forms of ranch management in similar environments (Cossins 1985; de Ridder and Wagenaar 1986). It is clear that pastoralism is potentially an economically viable and sustainable form of land use in the savanna lands of Africa.

Pastoral systems, however, are under stress in many areas: through war and conflict, through changing market structures, through environmental change and through the removal of land resources from pastoral production.

This paper examines pastoralists' ecological management strategies with a case study from the Barabaig of Hanang district, Tanzania. By providing details of the traditional Barabaig pastoral land-use system, it will be argued that the way the Barabaig manage rangeland resources is a rational and sustainable form of land use. Evidence is given to show that their common-land tenure arrangements are both sophisticated and effective for both production and conservation of land but that an inappropriate and costly development project has undermined this system.

The case study illustrates the rationale for traditional Barabaig natural resource management. Such management is based on a close understanding of the interactions between components of the landscape and the requirements of a range of resources to sustain production. The case study demonstrates how an inappropriate assessment of the economic value of the pastoral production system, and the resources that it depends upon, has led to external intervention that undermines the sustainability of the existing system. The chapter argues for the development of a more thorough understanding of the dynamics of pastoral production systems and livestock-resource interactions. This would allow more effective support for land tenure rights and common-property management systems.

PERSISTING WITH OLD ORTHODOXY

The plight of the Barabaig is typical of a wider problem for pastoralists throughout Africa. Common-land tenure systems have long been thought incapable of efficient land use (Lane and Swift 1989). Evidence of rangeland erosion since colonial times has been blamed on the "irrational" behaviour of pastoralists to accumulate cattle for the reasons of social prestige in excess of their economic needs. Credence was given to this view by reference to the "cattle complex" that dwelt on the cultural importance of cattle in traditional pastoral societies (Herskovits 1926). This anthropological legacy has since been reinforced by the "tragedy of the commons" scenario which posits that individual herders have no incentive to restrict stock numbers, and the herding of private animals on communal pastures will inevitably lead to overgrazing and land degradation (Hardin 1968).

Despite the years that have elapsed since Hardin first published his essay more than 20 years ago, and despite a mounting challenge to it on both theoretical and empirical grounds (Runge 1986), it has remained a powerful force in the minds of government officials and aid agency personnel to this day. A major problem has been the failure to distinguish between "common property" regimes with traditional control of individual exploitation and "free" or "open" access where such controls do not exist. This is highlighted in the continued espousing of the "old orthodoxy" that pastoralists have too many cattle and that, because of uncontrolled use of the commons, they will inevitably destroy the land through overgrazing (Lane and Swift 1989).

Despite the reality of pastoral existence in Tanzania, new thinking on pastoral development is not accepted by planners. An almost verbatim expression of the "tragedy of the commons" scenario is found in an article detailing measures for the "proper management of rangelands" by an official of the Tanzanian Ministry of Agriculture and Livestock Development:

> ... this practice of grazing private livestock on communal land constitutes the single major constraint to improved management of the natural pasture lands. The inevitable result of this system of livestock production is that the cattle owners keep excessive numbers of livestock which in turn leads to overgrazing, soil degradation, low fertility and high mortality rates (Bilali 1989).

Despite Tanzania's ranking as one of Africa's highest cattle resource countries, livestock production has failed to satisfy national demand for beef or dairy products and failed to exploit potential export markets (Raikes 1981). In response to this the government in 1982 published a National Livestock Development Policy that remains a guide for the direction of development in the livestock sector to the present day (Tanzania Government 1982). The preamble acknowledges that 99 per cent of livestock were in the hands of traditional producers, and that there was enormous potential for increased production. Yet the measures given to convert this wealth are based on the transformation of traditional pastoralism because it is regarded as backward and unproductive:

> The long term objective is therefore to bring about changes in traditional producers attitude and practices thereby increasing productivity to the level where this sector evolves into a modern subsector (1982:4).

The "modern subsector" envisaged in this context is made up of private and state ranches that do not include much pastoral involvement (Mustafa 1986). This has meant appropriation of large tracts of land and the destruction of pastoralists' traditional land-use management systems based on common land tenure.

95

THE BARABAIG OF THE HANANG PLAINS

The Barabaig are semi-nomadic Nilotic pastoralists who are a subtribe or section of a wider ethnic grouping known as the Tatoga. They number more than 30,000 in Hanang district. They fled from an invasion of the more numerous and powerful Maasai and surrendered occupation of the Serengeti plains and Ngorongoro highlands more than 150 years ago (Saitoti 1986). They moved south along the Rift Valley as far as Singida before returning to what they now regard as their territory on the Hanang plains (Wilson 1952).

All Barabaig herders strive to be self-sufficient from production of their cattle. Each household head manages his herd to maximise production of milk, meat and occasionally blood. However, they do not exist on a purely pastoral diet. Only those with very large herds receive half or more of their food energy needs from cattle products (Lane 1991a). Maize makes an important contribution to the nutrition of all Barabaig and is especially important in poorer households with fewer cattle. Grain is obtained through exchange or sale of livestock, and from shifting cultivation by households with the help of communal labour provided by relatives and neighbours.

Many studies have shown that pastoralists do not have excessive numbers of livestock. Even so-called "pure pastoralists" who practise a minimum of cultivation rarely have enough livestock on which to subsist on a pastoral diet (Swift et al. 1989). The average total number of cattle in a Barabaig household at Balangda is only 60 animals, or six per person. As only 30 per cent of these are milk cows, it means that each person receives milk from only two cows. As Barabaig cows provide only just over half a litre of milk a day for human consumption and their lactation is only for around 7–8 months, current herd size is clearly not enough for everyone to be self-sufficient in milk throughout the year (Lane 1991b). That is why the Barabaig and other Tanzanian pastoralists cultivate crops to supplement their diet.

The Hanang plains are divided by the Great Rift Valley escarpment which separates the northern and elevated Basotu plains from the lowland Barabaig plains to the south (see Figure 6.1). The Basotu plains are undulating with a series of depressions and many low hills and volcanic craters. As there are no perennial rivers, crater lakes provide some of the few sources of permanent water. Even with the provision of dams, built in colonial times, however, there is a severe shortage of water in the dry season.

The Barabaig plains, on the other hand, have permanent water from lakes Balangda Gidaghan and Balangda Lelu. These lakes also provide salt for the Barabaig and their livestock. Mount Hanang dominates the landscape rising to a peak of 3118 m above the surrounding Barabaig

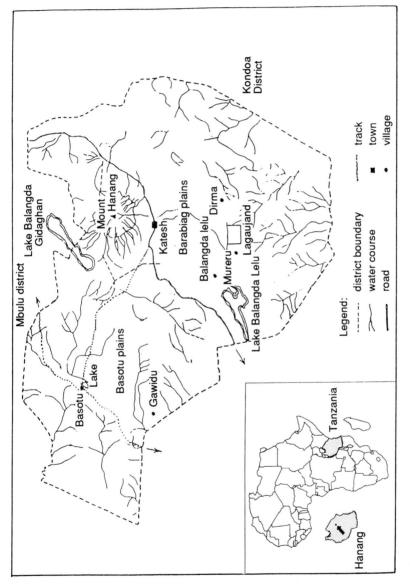

Figure 6.1 Map and location of Hanang district. Source: Agriteam Canada

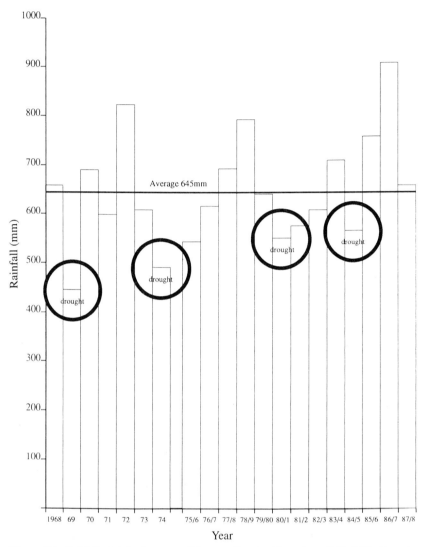

Figure 6.2 Rainfall distribution over 20 years, Basotu Plantation 1968–1988. Drought periods are circled

plains. The plains extend to the south as far as the Mureru range. Apart from bushland beyond the range, and the forest of Mount Hanang, the dominant vegetation is *Acacia* and *Commiphora* woodland interspersed with open grassland. The climate is semi-arid with periodic droughts and an average annual rainfall of around 600 mm (see Figure 6.2).

Hanang district can be divided into several ecological zones with contrasting soil and vegetation. Much of the area is covered with calcareous volcanic tuff, spread over the underlying granitic shield by the eruption of Mount Hanang. In the Basotu and Barabaig plains areas, soils are derived from volcanic materials and some are of very high fertility. To the south, sandier, poorer fertility soils are dominant, derived from granitic base rock supporting miombo woodland. In the volcanic soil areas, vegetation is characterised by a mixed *Acacia–Commiphora* woodland with areas of open grassland. The rift escarpments divide two major plains areas, establishing an important boundary within the district (Fenger et al. 1986; Newman 1970).

Three major zones can be identified:

(1)   The volcanic, high fertility Basotu plains;

(2)   The lower altitude volcanic soil Barabaig plains; and

(3)   The sandy soil miombo bushlands to the south.

The major zones can be differentiated according to soil fertility characteristics and in relation to underlying geology. In savanna systems these properties have a major impact on both primary and secondary production dynamics (Frost et al. 1986).

Within each zone there are patches of distinct environments. These patches differ in size. They include the vertisol soil depressions of the volcanic plains which constitute a significant proportion of the Basotu plains; the riverine vegetation strips associated with the rivers and streams draining the higher ground; the lake margins of the crater and salt lakes, notably Lake Balangda Gidaghan and Lake Balangda Lelu, the valley bottomlands (*mbuga* in Kiswahili) of the bushland miombo zone; and, at a smaller scale, the particular vegetation types associated with abandoned homestead sites. Although small in area, these may represent "key resource" patches in terms of livestock production. Understanding the contrasting ecological dynamics of these zones, and evaluating the role of patches within them, provides the basis for interpreting Barabaig natural resource management strategies (Scoones 1991).

Fenger et al. (1986) record some basic soil characteristics of the major zones and patches within them. Soils in three areas can be contrasted (see Table 6.1).

Tables 6.2a and b extend the analysis by illustrating the range of resources according to the Barabaig classification of resource types. Soil type and underlying geology are related to availability of fodder species (grass and browse) and the key constraints to livestock production in each zone or patch. The use of the Barabaig ecological classification allows the interpretation of ecological pattern according to local criteria.

Table 6.1 Soil characteristics of different zones and patches at approximately 25 cm depth. CEC = cation exchange capacity. Source: Fenger et al. (1986)

Sample area	Clay content (%)	pH	CEC (meq/100 g)	P (ppm)	K (meq/100 g)
Basotu plain (Topland, volcanic soils; mollisols)	50	6	35	50	3
Basotu plain depressions (vertisols)	65	8	80	5	1.75
Rift valley/boundary area with granite soil bushland	30	6	15	1	1

Different zones within the district have contrasting ecological dynamics and constraints to pastoral production. The grasslands of the high fertility, volcanic soil plains have high potential productivity and provide good quality fodder. Variability in rainfall levels means that primary production is characterised by instability. Although these eutrophic savannas can potentially support high populations of livestock (Bell 1982; Coe et al. 1976), constraints of water availability mean that pastoral production must rely on "key resource" areas and flexible drought responses. The depression areas (*muhajega*; sing. *muyaded*) of the Basuto plains provide important wet season fodder, while the river banks and lake margins of the Rift Valley provide dry season fodder refuges. In times of drought, movement out of the area into neighbouring bushland and river valleys to the east and south of Hanang district sustains livestock production.

The bushlands to the south present a different dynamic pattern. Inherent low soil fertility of the dystrophic miombo savanna type predicts lower potential stocking rates according to Bell (1982). Although of lower biomass productivity and quality, fodder production can be expected to be more stable over time in this savanna type, with stability being increased by the availability of key bottomland grazing resources (e.g. *mbuga* depressions). The incidence of tsetse in these woodlands provides a further constraint to livestock production.

Barabaig livestock production is thus characterised by a tension between drought and lack of permanent water sources in the potentially highly productive volcanic soil plains areas, and tsetse incidence and savanna areas of lower pastoral potential to the south (cf. Birley 1982). The use of the area for livestock production will be centrally determined by trade-offs between the productivity and stability of grassland production in each zone and the existence of the different constraints of water availability and tsetse

Table 6.2a Landscape components and livestock production potentials and constraints in the *darorajand*, *muhajega* and *darorajega* used by the Barabaig

Land type	Geology	Soils	Tree spp.	Grass spp.	Constraints
Basotu plains (*darorajand*)	Calcareous volcanic tuff	Mollisols Deep fertile	*Acacia* spp. *Commiphora* spp. *Grewia* spp.	*Pennisetum* sp. *Aristida* spp. *Setaria* spp. *Heteropogon* sp. *Hypparhenia* sp.	Lack of available water in dry season. Grass failure in droughts.
Bottomlands (*muhajega*)	Volcanic tuff	Vertisols Deep fertile	Few trees *A. drepanclobium*	*Eragrostis superba* *Pennisetum mezianum* *Cynodon dactylon*	Appropriation for wheat farms. Gully erosion in farms.
Barabaig plains (*darorajega*)	Volcanic tuff	Mollisols Deep fertile	*Acacia* spp. *Commiphora* spp.	*Pennisetum* sp. *Aristida* sp.	Intensity of use. Lack of grass in poor rainfall.

101

Table 6.2b Landscape components and livestock production potentials and constraints in the *ghutend*, *hayed*, *darabet* and *mbuga* used by the Barabaig

Land type	Geology	Soils	Tree spp.	Grass spp.	Constraints
Bottomlands: river and lake margins (*ghutend*)	Fluvial and lacustrine deposits	Sandy/silty clay loam Salt pans	Few trees	*Chloris gayana* *Cynodon dactylon*	Heavy livestock pressure Salinity
Hills/mountains escarpments/ ridges (*hayed*)	Volcanic tuff (plus granitic intrusions)	Lithic thin, stony, fertile clay loam	*Acacia* spp.	*Aristida* spp.	Mt Hanang forest reserve Steep slopes
Miombo woodland (*darabet*)	Granitic	Loamy sands Sandy loams infertile	*Brachystegia spiciformis* and other miombo spp.	*Aristida* spp. *Brachiaria* spp. *Tragus* sp. *Dactyloctenium* sp.	Tsetse fly Poor forage quality
Valleys and depressions (*mbuga*)	Granitic	Colluvial heavy fertile	*Cassia* sp. *Commiphora* sp. thickets	*Brachiaria* sp. *Dactyloctenium* sp.	Tsetse fly

incidence. The sustainability of the pastoral system over time is critically dependent on a flexible response to the changing patterns of resource availability and production constraints. This requires an opportunistic mode of natural resource use that allows for the exploitation of "key resource" patches, such as *muhajega/mbuga* and river or lake margins, at particular periods. These responses must be complemented by movement responses during drought periods.

Flexible movement, on a seasonal basis within the district and, in some years, outside the area, is thus the key to population persistence and sustainability. Wildlife movement in response to variations in fodder quantity, quality and mineral nutrition has been extensively examined for savanna in the Serengeti (e.g. Sinclair and Norton-Griffiths 1979; McNaughton 1985, 1990). A clearer understanding of similar patterns of ecology in other pastoral livestock populations is also emerging (Breman and de Wit 1983 for the Sahel; Scoones 1990 for Zimbabwe).

BARABAIG MANAGEMENT OF SAVANNA RESOURCES

The following sections discuss the key elements of the Barabaig natural resource management strategy.

Traditional seasonal grazing rotation

Variety of soil types, topography, vegetation, and the availability of ground water provide the basis on which the Barabaig classify pastures (see Table 6.2). They recognise specific forage resources that are associated with eight geographical features. So as to utilise pasture when it is most productive, and to rest areas and allow them to recover, the Barabaig have devised a seasonal grazing rotation that exploits the forage regimes at different times of the year. An opportunistic response during times of drought is particularly important.

The rotation is not always regular as climatic variation can cause deviations (see Figure 6.3). Presented here is a stylised model of a complex and variable cycle.

The Barabaig year begins in May. This is the start of the late rains (*mehod*) when pasture availability and livestock production are at a maximum. It is a time of relative plenty and recovery from the deprivations of the dry season. It is also the time when the Barabaig like their herds to be on the *muhajega*. These are depressions on the plains containing fertile soils that sustain a mix of grasses and herbs that the Barabaig call *nyega nyatk*.

The Barabaig regard the forage of the *muhajega* as the most productive pasture available to them. It is valued for its capacity to produce high milk

103

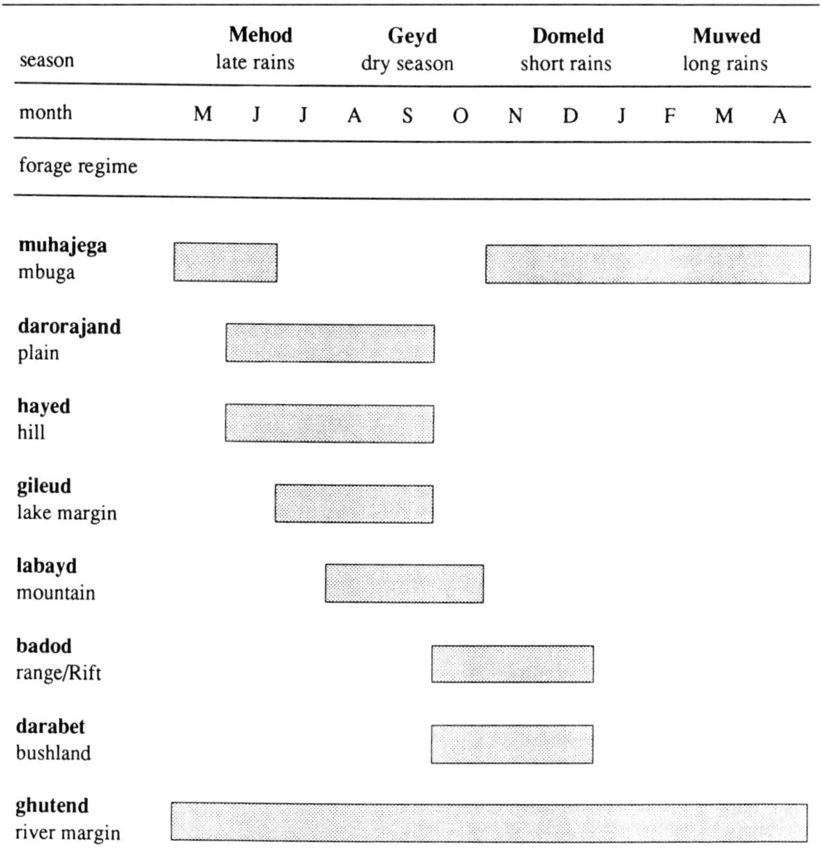

season	Mehod late rains			Geyd dry season			Domeld short rains			Muwed long rains		
month	M	J	J	A	S	O	N	D	J	F	M	A

forage regime

muhajega mbuga												
darorajand plain												
hayed hill												
gileud lake margin												
labayd mountain												
badod range/Rift												
darabet bushland												
ghutend river margin												

Figure 6.3 Traditional Barabaig grazing rotation

yields, induce growth, and for its recuperative powers for livestock suffering ill health from the stresses of the dry season and droughts. The *muhajega* are found in greatest abundance on the Basotu plains. It is on these plains that the Barabaig congregate during the rains. This is only possible for as long as surface water is available there.

When the rains stop in May, June or July, and surface water dries up on the Basotu plains, many Barabaig move down the Rift escarpment onto the Barabaig plains. Here they are able to draw on permanent water from wells at Lake Balangda Lelu that sustain them through to the months of August, September and October of the dry season (*geyd*). At this time they gather on the plains (*darorajega*) of Balangda and the many low hills (*hayed*) to the east of the lake. As grazing becomes depleted they begin to move to the miombo bushland (*darabet*) with its associated risks from tsetse flies and predation from wild animals.

As soon as the short rains (*domeld*) begin in November or December and surface water is found again on the Basotu plains, the Barabaig move their herds back onto the *muhajega*. They remain there for the 6 or more months of the rainy season (*domeld, muwed* and *mehod*) until the water dries up again. But for the river margin (*ghutend*), the *muhajega* supports the Barabaig for longer than any other forage regime. As the *ghutend* is confined to the few small perennial rivers that flow a short distance from Mount Hanang it is much less important than the *muhajega*, which the Barabaig value more than any other forage resource.

Thus, for the Barabaig, availability of water acts as the most limiting factor in pasture use. Whereas most other pastoralists need to locate pastures in the dry season to sustain them (sometimes they are preserved for this purpose), the Barabaig are forced to move away from their richest forage regime when they need it most. The availability of dry season "key resources" on the lake and river margins, however, are critical during this period. These areas must be complemented by the plains grassland of the *darorajega*, where in the dry season animals congregate near the permanent water afforded by wells at Lake Balangda Lelu.

Grazing management

Grazing on the plains areas is also actively managed. Deliberate burning of pasture is called *ghwardaida ngy'anyid* and is mainly practised between September and November to burn off remaining dry stalks of vegetation that have survived the dry season. This allows rapid vegetative response of new shoots with the first rains. This is preceded by *lajitaghed*, the burning of firebreaks around the homesteads. Sometimes such fires do get out of control, but every effort is made to manage fire carefully lest valuable dry season fodder is destroyed.

The Barabaig not only use burning as a management tool for grass production, they also use it as a strategy for reducing the incidence of ticks. Tick-borne diseases are a major cause of livestock mortality in the area (Lane 1988). Burning, combined with intentional heavy grazing, helps to reduce the extent of tall grass stands and so tick populations.

Tsetse control measures

Tsetse incidence in the miombo bushlands is a major factor preventing the exploitation of this area. Dry season pressure on other grazing resources means that herds are often forced to enter this zone. A variety of strategies is used to reduce tsetse fly populations. Burning intentional heavy grazing both act to reduce tsetse habitats and to increase the quality of grazing in the poor sandy soil areas. Bush clearance is another strategy for improving the pastoral potential of this savanna type. The adoption of a shifting cultivation system by the Barabaig removes dense bush cover in certain areas. Cultivation is concentrated at Mureru along the ecotone between the Barabaig plains and the miombo bushland. As the Barabaig remove the bush to cultivate, they are not only reducing the area of tsetse infestation, but also expanding the area for dry season grazing.

Controlling resource access: common-property management

Barabaig movement of herds and homesteads around the plains is a response to the vagaries of climate, so as to make best use of scarce and variable pasture resources. To make this possible the Barabaig accept that everyone must have general access to pastures. Open rangeland is regarded as the property of the whole community. This enables herders to choose when and where to be on the plains according to their individual assessment of pasture quality.

So as to facilitate this use of forage resources the Barabaig, like other African pastoralists, have a form of common-land tenure (see, for example, Makec 1988). They describe the land that makes up the range as *ng'yanyida madagh* (*madagh* = common). But this is not a case of universal and uncontrolled access to land and its resources by everyone and anyone. The Barabaig recognise the right to protect land; *weta ng'yanyid* (*weta* = protect). In the past, they also recognised the need to protect this resource from intruders and deployed a warrior set to act as a guard. Due to the breakdown of the process that formed this socio-political unit, and as a result of the imposition of centralised control by the state, however, they are no longer able to defend their territory as they had done before.

Within the commons, rights to property range from communal access to exclusive private property. The Barabaig regulate rights of use and access to land through a tripartite jural structure, each with its own sphere of interest and authority: the community, the clan, and the individual household. In this system access to and use of land is controlled by a set of customary rules.

The Barabaig have a hierarchy of jural institutions that control access to and use of land, interpret customary rules and adjudicate in rare conflicts over rights and duties. Matters concerning the community as a whole are dealt with by a *getabaraku* or public assembly of all adult males. This has ultimate authority on common-property rights over open rangeland. Decisions are made by consensus. If the problem has particular gravity or involves the application of sanction, a committee of the *getabaraku*, called *makchamed*, is formed. The *makchamed* deliberates in camera and hands down its decision to the assembly.

Women have their own council (*girgwageda gademg*) which has recently involved itself with land issues. Their interest in land stems from their authority on matters involving offences by men against women, and their special role in Barabaig spiritual life. Desecration of sacred property or the failure of men to protect such property is currently of particular interest to them.

At the clan level, a clan council (*hulandosht*) controls all clan property, such as burial sites (*bung'eding* (pl.)) and wells. At the neighbourhood level, authority is vested in a neighbourhood council (*girgwageda gisjeud*). Anyone wishing to come and live in a neighbourhood can only do so with the endorsement of this council. In this way the council effectively controls the entry of too large or too many herds and limits the potential for overgrazing.

Sanctions against those who infringe common-property rules generally take the form of an order from whatever judicial moot has authority in the case. Most often the offender is simply asked to desist from the offence. If damage is done, an offender may be asked to pay a fine to the elders, or compensate an individual who has suffered as a consequence of the offence. Fines are collected by a group of youths (*orjorda*) which ensures that the transgressor complies with the ruling. Punishment for more serious offences can be greater. In the past the women's council has been known to have issued a curse (*moshtaida*) on male offenders for breach of rules related to sacred land. This curse is most effective as it is known to bring ruin to people's lives and is much feared by all Barabaig (Lane 1991a).

Grazing resource use is controlled through access to water; by regulating use of a key limiting factor, an efficient system of control can be evolved. Although surface water is universally accessible to everyone, its use is controlled by rules. Routes to and from water are not to be restricted by the construction of homesteads, and water sources must not be diverted or contaminated. A well becomes the property of the clan of the man who digs it. Although anyone may draw water from a well for domestic purposes, only clan members may water their stock there.

Rights to trees are also subject to certain rules. The Barabaig protect sacred trees, such as the *Ficus* species, by banning any form of use or damage to them. Other trees that are regularly used for meetings (predominantly *Acacia* spp.) are also protected in this way. Clans also have rights that cover property in land. The most important example of this is the burial site (*bung'ed*). Private property is recognised by the Barabaig in the form of a man's homestead and its surrounds. For example, a grazing refuge (*radaneda nyega*) is marked off with thorn tree branches to protect forage near the homestead, which can support small stock and sick animals that cannot keep up with the main herd. Private property also extends to trees that give shade to the household and those selected to hang beehives.

Effective common-property management is based on regulation of access to areas that are enforced through sanctions (Runge 1986). The Barabaig system of resource control includes such components. It is far from a pattern of open access assumed by the "tragedy of the commons" scenario. To remain effective, however, common-property management systems must be based on definable rights supported by legal instruments to ensure that the system is not undermined (see below).

Regulations to control degradation

Barabaig regulations to control resource access also aim to limit degradation, particularly of key resource areas. This is mediated through customary regulations on the siting of settlements.

Without adherence to the movement of herds and rotation of pastures, some herders would possibly remain at the river margin (*ghutend*) where there is permanent water and persistent vegetation. The Barabaig, however, understand that this would result in destruction of this resource and have a customary rule that bans settlement at the *ghutend*, and denies herders the right to graze the forage if they are not there to water their stock.

As water in the crater lakes of the Basotu plains is insufficient in the dry season to sustain all the livestock that congregate there, some herds have to be moved elsewhere. The wells of Lake Balangda Lelu provide alternative permanent water. Their use, however, is dependent on dry season fodder availability in the surrounding plains. Like restricted use of the *ghutend*, grazing of the Balangda plains must be limited if pasture is to be available at the time when it is most needed. To achieve this the Barabaig traditionally banned permanent habitation on the *darorajega* at Balangda. This customary rule was made a by-law by Gejar, Chief of the Barabaig Native Authority (*ca.*1936–1952), who recognised the threat that permanent habitation on the *darorajega* would have on this vital resource.

BARABAIG ECOLOGICAL MANAGEMENT: A CHALLENGE TO THE OLD ORTHODOXY

The components of Barabaig natural resource management strategies outlined above represent a challenge to the old orthodoxy. It is clear that resource use is not based simply on an "open-access" situation, but regulations control use under a common-property management system. Access to a range of resources within different ecological zones and savanna types is critical to the sustainability of the system. Allocation of land title in restricted blocks would disrupt patterns of flexible movement that characterise the pastoral system.

Range management recommendations that demand destocking, restriction of movement and the cessation of grassland burning or shifting cultivation in order to "conserve" rangelands from "degradation" can also be seen to be misconstrued. High stocking rates are required to sustain the economic livelihoods of the population and also are used as an explicit strategy to control disease incidence. Controls on movement would also undermine the key pastoral strategies to ensure survival, during droughts and resource shortage periods during the year. Burning of grasslands and bush clearance can also be interpreted in terms of a basic pastoral rationale to offset the dual threats of the tick and tsetse fly. Analysis of Barabaig natural resource management strategies therefore suggests the need for a re-evaluation of "orthodox" range management recommendations.

There are, however, limits to the ability of traditional Barabaig practices to cope with changing circumstances. For instance, with increased land pressure from encroaching Iraqw and Nyaturu farmers, some Barabaig have had to ignore the customary rule and live permanently on the Balangda *darorajand*. Today, neither customary rules nor new state laws have protected the *darorajand*, creating a legal vacuum and a situation of virtual open access. Similarly, removal of access to *muhajega* grazing resources through the establishment of commercial wheat farms in the Basotu plains has increased pressure on other resources, resulting in land degradation (see below). Conflicts arise during drought periods when migrations are made outside the area in search of fodder. Movement options are restricted because of the existence of extensive agricultural lands in adjacent districts.

Such contemporary constraints are not arguments for abandoning the pastoral land-use system, but should prompt strategies for reinforcing its central components and addressing issues of conflict. This requires a reassessment of the old orthodoxy and an exploration of patterns of technical and legal support for common-property management.

The appropriation of land for agriculture has been the most serious threat to the Barabaig production system. The consequences of loss of land due to the establishment of a large-scale wheat project is explored below.

LOSS OF LAND: TANZANIA–CANADA WHEAT PROGRAMME

Since the beginning of the colonial era the population of Hanang district has increased dramatically. Human pressure on land in the Mbulu highlands to the north of the district resulted in Iraqw farmers encroaching on what was Barabaig grazing land. Although the British administration attempted to stop this, it proved impossible (Schultz 1971). After a brief respite from encroachment by way of the Barabaig Native Authority prior to independence, encroachment on Barabaig pasture land continues to the present day.

The practice of a seasonal grazing rotation means that at certain times of the year, and sometimes for long periods, land is free of human habitation and grazing livestock. This has led some people into thinking that some pasture land is unoccupied. It is because of this that Barabaig pastures are sometimes taken over by others and put to non-pastoral use.

The inherent fertility of the volcanic soils of the Basotu plains makes it ideal for farming. Inevitably, farmers take the best agricultural land. Just as this land has attracted Iraqw farmers, so its agricultural potential has also been noticed by developers. In 1968 a Canadian mission identified many of the soils on the Basotu plains as "highly suitable for mechanised dryland farming" (Fenger et al. 1986). This has led to government appropriation of large tracts of land on the plains, for an extensive wheat scheme. In the early years of the scheme the critical grazing resource of the *muhajega* was not regarded as ideal for mechanised agriculture, because of the dry season cracking and wet season waterlogging of the vertisols. In the later stages of the scheme development, however, these areas were also included and "land suitability" assessments revised (MacMillan and Ngoma 1983).

In response to an expected increase in demand for wheat, the Tanzanian and Canadian governments agreed to develop the Basotu plains for agricultural production via a Tanzania–Canada Wheat Programme (TCWP). The scheme is based on a technological package reminiscent of the "green revolution", with mono-cropping of hybrid varieties using large-scale mechanical equipment along the lines used by prairie wheat farmers in Canada (Lane and Pretty 1990). The Tanzanian partner in the venture – National Agriculture and Food Corporation (NAFCO) – set about acquiring land in the area for seven farms of 10,000 acres each, making a total combined area of 70,000 acres for the scheme. This was later increased to 100,000 acres.

The TCWP farms now cover 12 per cent of all the Hanang district land (see Figure 6.4) (Lane and Pretty 1990). The significance of this loss to the Barabaig, however, is far greater than just the land area involved. Together with the land occupied by encroaching farmers, it represents the loss of virtually all the crucial *muhajega* forage resource that is so important for pastoral production. This has altered the seasonal grazing rotation and had a negative effect on herd productivity and Barabaig welfare (Lane 1991b).

Despite the fact that some Barabaig were resident in the area appropriated for the wheat scheme and that the *muhajega* were a vital forage resource,

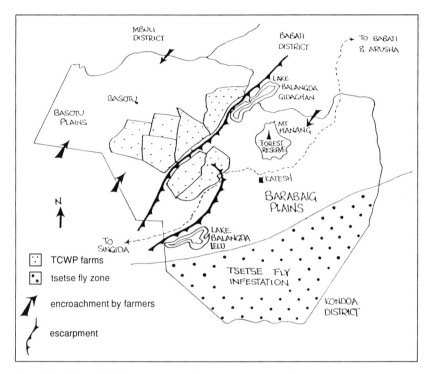

TCWP farms

tsetse fly zone

encroachment by farmers

escarpment

Figure 6.4 Map of the Hanang district showing major land-use features (TCWP = Tanzania-Canada Wheat Programme)

this land was described as "idle" during project assessments (Young 1983).

Some serious inadequacies in the economic evaluation of the wheat project are highlighted on closer examination.

THE ECONOMIC VALUE OF PASTORAL RESOURCES

The wheat project assessment was inadequate on two important counts. It overestimated the potential economic returns of the scheme and ignored the opportunity costs of reductions in land to the pastoral system.

Assessing the scheme's economic performance

The economic performance of the TCWP has been challenged by both internal evaluations and independent assessments. Early evaluations of the scheme showed positive cost/benefit ratios and a 40 per cent internal rate of return (Stone 1982). In 1982, however, the economic realities of the TCWP came to light. Two internal evaluations revealed that, whilst the farm accounts showed the TCWP was profitable in financial terms, the full economic costs including aid, subsidy and tariffs protection exceeded the economic benefits (Prairie Horizons 1986; Michael Mascall and Associates 1986). In addition, increasing problems of soil erosion in the farms (Fenger et al. 1986) highlighted the additional "hidden" environmental costs of the scheme; these never entered the economists' equations.

In 1986 an independent study was done with the purpose of testing the efficiency of resource use and to find out whether it makes economic sense to grow wheat using the methods employed on the scheme when compared with smallholders using oxen, and relative to the cost of direct importation of wheat from the rest of the world market (Carter et al. 1989). The study concluded that the scheme had a negative net financial profitability and a negative cost/benefit ratio. This led to the conclusion that the TCWP "is not economically viable nor does it make effective use of domestic resources in saving foreign exchange for Tanzania to use large-mechanised wheat production to satisfy the domestic demand for wheat" (Carter et al. 1989).

Opportunity costs to the pastoral system

These re-assessments of return from the wheat project must be combined with losses in economic returns to the pastoral system. Assessing the economic value of pastoral resources is complex. Costs and benefits are difficult to assess in quantitative terms, presenting problems of subsistence production valuation and the assessment of the costs of environmental impacts. Difficulties in economic evaluation are also compounded because use is made of different areas at different times of year. Benefit streams from different components of the landscape thus vary over time, both seasonally and interannually. Land is not "idle" if it is not used continuously. Indeed, access to particular key resources may be critical to the long-term sustainability of the system. Removal of land from pastoral use may impose major economic costs in the long term, both through reduced pastoral production viability and increased land pressure and environmental costs on other areas.

A framework for economic assessment

Such complexities, however, are not an excuse for ignoring these issues completely. Recent developments in environmental economics suggest possible frameworks for economic valuation in pastoral situations (Barbier

1989; Dixon et al. 1990). Box 6.1 outlines the possible components of such an economic framework that would be needed for a more complete evaluation of the direct costs of removing key grazing areas from the Barabaig and placing them into the TCWP. Box 6.2 then presents information on the *indirect* and *non-use* costs associated with this loss.

The loss of land to the wheat scheme can be seen to have a significant cost. A quantitative assessment is impossible with currently available data; however, a framework for assessment is proposed. A critical point is that the opportunity cost of lost *muhajega* land is not simply the reduced value in proportion to the area removed. This is because of its crucial role in the seasonal grazing rotation. The economic value of key resource patches needs to take a central place in economic assessments. The impacts of this land loss are considered further below.

IMPACT OF LAND LOSS ON THE SEASONAL GRAZING ROTATION

The impact on the seasonal grazing rotation by the loss of *muhajega* has forced the Barabaig to adopt a new grazing pattern (see Figure 6.5) and to rely more heavily on the seven remaining forage areas.

They have been forced to use of *darorajand, hayed*, and *gileud* forage when these would otherwise be rested from grazing. Further, together with the *ghutend*, it means that these areas are more intensely used at traditional times of use. Only the mountain (*labayd*), Rift escarpment (*badod*) and bushland (*darabet*) are relatively unaffected by the loss of the *muhajega*, as these are unable to match the productivity of the other forage regimes. Shifts of resource use to these low resource potential areas, however, may result in adverse impacts on the environment and animal productivity. Possibilities include more erosion as a result of increased use of shallow soil hill slopes and declines in animal productivity from increased exposure to tsetse fly in the bushland.

Environmental impacts: degradation

Changes in pastoral land use induced by the establishment of the wheat scheme have resulted in a range of environmental impacts. These are outlined in Table 6.3. This is based on information derived from data collected and interviews with the Barabaig carried out during 1987 (Lane 1991a), plus environmental assessments carried out in the wheat scheme area (Fenger et al. 1986). Impacts are not uniform and it is necessary to dissect the range of impact in different land types and explore the consequences for livestock production in the area.

The land resources of Hanang district support a pastoral population of approximately 35,000 Barabaig with around 207,500 cattle (based on a 1984 census with 2.25 per cent annual growth). This represents a considerable standing capital value. If each animal is conservatively valued at $US 84 (c.Tsh. 8,377/- actual average sale price for marketed animals in 1987) this represents a total value of approximately $US 17.5 m. This price does not include premiums that would be paid for productive females (as high as Tsh. 30,000/-, c.$US 300 per animal). The flows of products from these livestock include the regular production of meat and milk. Assuming a conservative 5 per cent meat off-take this represents an annual flow of $US 726,000. Milk products should be added. Assuming production of 157.5 litres/lactation are available for home consumption, a calving interval of 15 months and 30 per cent of the total herd as milch cows, this results in a total yield of 4.57 million litres per annum (Lane 1991). If milk is valued at 50 US cents per litre (the average replacement cost of milk purchased in Babati in 1987; range: 35–60/-), the total value of annual milk production is $US 2.3 m.

A full assessment of herd value should also include the increments to the herd through births and acquisition (approximately 2 per cent per annum). Based on calculations of milk and meat outputs only, the "average" animal produces approximately $US 14.5 worth of output per year. Any significant reductions in herd size or decreases in productivity would thus result in a significant cost.

Further outputs from the system include wood products, agricultural produce, wild foods, and honey. A total assessment of direct use value would have to consider the whole range of products harvested from the savanna ecosystem.

Estimated costs of the removal of land by the wheat scheme are difficult to assess. However, declines in cattle populations in areas adjacent to the farms have been assessed by Barabaig elders at around 30 per cent over the past 7 years. Productivity levels of the remaining herd have also been reported to have declined due to the loss of land. If a third of half the district herd is depleted by the scheme then there is an additional capital loss of $US 2.9 m.

Such economic losses due to decreased livestock populations need to be reflected in the assessment of the direct costs of the scheme; these are likely to be significant if the approximate valuations attempted above are an indication.

Box 6.1 A Barabaig perspective of the direct costs and benefits of the TCWP

Coupled with the direct costs and benefits of key savanna areas to the Barabaig, it is necessary although difficult, to assess the economic value of the ecological functions of the various components within the landscape. Soil and water processes in savanna ecosystems remain poorly understood. Rigorous assessment of the economic benefits of such processes to production systems is therefore impossible. However, the impact on ecological functions should not be excluded from an economic framework that incorporates considerations of sustainability (Barbier 1989).

The impact of mechanised farming operations on the Basotu plains includes the costs of soil erosion. The long-term impact of this on the productivity of the system, through siltation or disruption of hydrological processes, remains unknown.

The Barabaig regard land as more than a physical resource. For them it can also have great spiritual significance. The Barabaig bury certain of their esteemed elders with a *bung'ed*, which is both the name of the burial mound and the funeral ceremony associated with it. The ceremony involves thousands of people and costs what amounts to a fortune for the Barabaig, including as it does the slaughter of many livestock and the brewing of vast quantities of honey beer. These burials constitute one of the most important expressions of Barabaig culture. The presence of a *bung'ed* provides an attachment to land that cannot be broken in Barabaig culture.

The ploughing up of burial sites as part of the wheat scheme clearly represents a major social cost of the scheme.

Box 6.2 A Barabaig perspective of the indirect costs and benefits of the TCWP

Various types of impact are identified in Table 6.3. Do these constitute "degradation"? Abel and Blaikie (1989) offer a definition of land degradation in pastoral systems that explicitly recognises the economic dimension. They describe degradation as "an effectively permanent decline in the rate at which the land yields livestock products under a given management system".

Degradation is thus measured in relation to the economic returns from the livestock enterprise. Changes in environmental factors therefore do not necessarily imply degradation; they must have a direct impact on the yield of livestock products and they must be irreversible. If this is accepted, certain impacts outlined in Table 6.3 may not represent degradation under this definition.

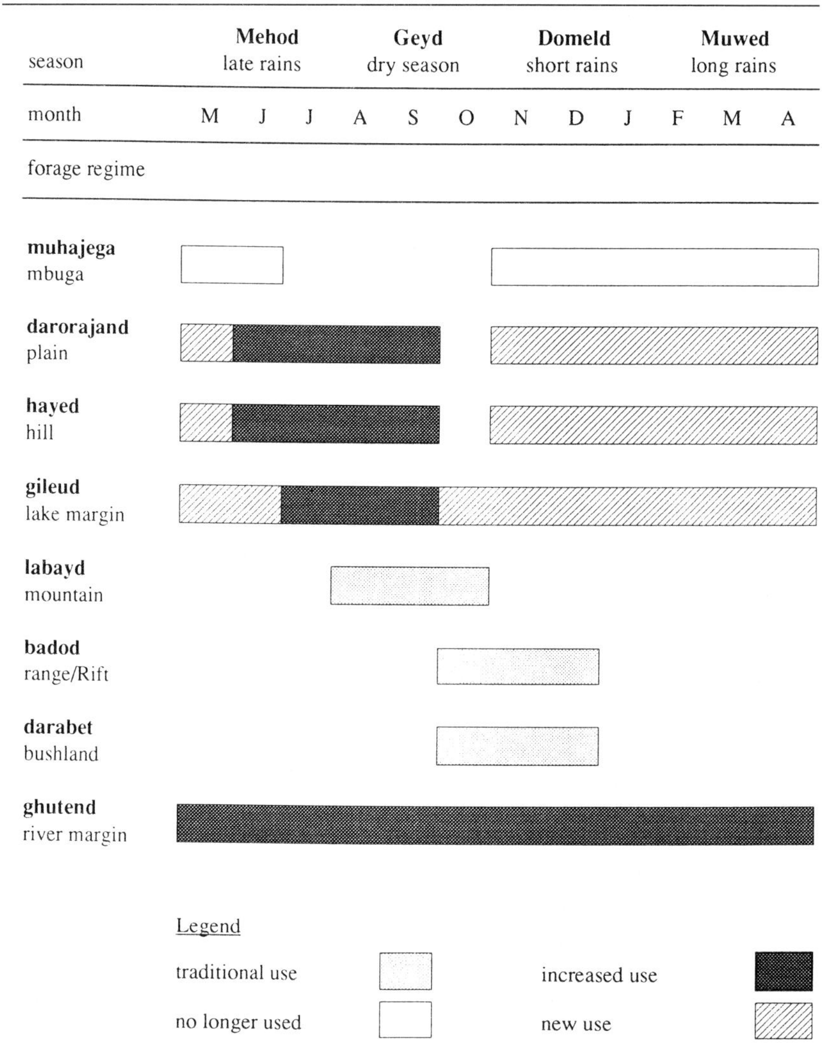

season	Mehod late rains		Geyd dry season		Domeld short rains		Muwed long rains		
month	M J J	A S O	N D J	F M A
forage regime				

muhajega
mbuga

darorajand
plain

hayed
hill

gileud
lake margin

labayd
mountain

badod
range/Rift

darabet
bushland

ghutend
river margin

Legend

| traditional use | | increased use | |
| no longer used | | new use | |

Figure 6.5 New Barabaig grazing pattern

For instance, loss of soil from hillslopes may result in longer-term reductions in grassland productivity, although the time scale over which this is likely to occur is unknown. The fact that livestock make little use of these areas suggests that this is probably not a key area of concern to the Barabaig. The same applies to the reduction in tree cover in the miombo bushland. In fact, because of tsetse reduction, this impact is likely to have a positive impact on livestock production.

116

Table 6.3 Forms of environmental impact due to the Tanzania-Canada Wheat Programme (TCWP)

Resource type	Impact
Topland range resources	
Plains	Decreased perennials
Hills/mountains	Soil loss
Bushland	Bush clearance
Bottomland key resources	
Muhajega	Farm land – gully erosion
River and lake margins	Heavy grazing, erosion, hoof damage
Lakes/springs	Siltation

Shifts in grass species composition may be due to rainfall changes; a period of drought naturally increases the percentage annual cover in savanna grasslands. This may be reversed with increased rainfall or decreased livestock pressure. Annuals may also provide good quality fodder during the rains with a rapid response to rainfall onset. Changes from perennial to annual grasslands, therefore, do not necessarily represent degradation. In the Barabaig plains, however, the decreased perennial cover has persisted over some time. According to local informants, this loss in perennial cover is not the result of recent droughts, since field observations followed a period of high rainfall (see Figure 6.1). This shift is also more likely to be detrimental to livestock production than in other savanna areas, since the plains areas are used for dry season grazing. A good dry season grazing resource requires perennial grasses that persist throughout the year. Although possibly reversible (with changes in livestock pressure), there is a good reason to regard the changes on the *darorajand* as "degradation". Certainly this concern is echoed by the Barabaig who point to the loss of key grass species in recent times (Lane 1990).

The impacts that represent the major threat to livestock production are those situated in the "key resource" areas of the river and lake margins and the *muhajega*. The removal of the *muhajega* from pastoral production has resulted in a transformation of the seasonal grazing pattern (see Figure 6.5). Increased reliance on alternative key resource areas has resulted in localised increases in stocking rate in riverine areas, lake margins and *muhajega* not used for the wheat scheme which have resulted in changes in resource productivity of these areas.

A focussed approach to environmental impact, that recognises the relationships between natural resource base and its use, is needed. This requires an understanding of pastoralist management strategies and ecological classification. In this case, recognition of the "key resource" areas has allowed the assessment of environmental impact in a systems context.

CONCLUSIONS: THE WIDER IMPLICATIONS FOR SAVANNA LAND MANAGEMENT

What are the lessons that can be drawn from this case study for the management of pastoral systems in savanna environments? Several inter-related themes are highlighted:

(1) Future plans for savanna land use should be based on existing traditional forms of management, and include the involvement of local land users in the formulation of land-use plans.

(2) It is necessary to understand the rationale for indigenous pastoral land use. Pastoralists' ecological management strategies are generally founded on a detailed understanding of the dynamics of the savanna system and the constraints of livestock production in such environments. The old orthodoxy of range management recommendations must be re-assessed in this light.

(3) Pastoral systems are dependent on a range of different land resources used at different times and in relation to different constraints. It is essential to identify "key resource" patches and management strategies (notably movement) that sustain the system. Recognition of their role will improve economic assessments of the impact of the land removal and degradation processes.

(4) Frameworks are needed for economic valuation of pastoral systems that incorporate appropriate valuation of direct use, indirect use and non-use values. The impact of external interventions should be assessed, with an explicit assessment of the opportunity costs and environmental impacts of land appropriation from pastoral areas.

(5) The management of natural resources in pastoral systems is based on regulation as part of common-property tenure systems. The patterns of regulation need to be understood and reinforced by securing legal rights over land (particularly key resources) and providing regulated access to areas needed for drought survival.

The case of the Barabaig of Hanang district is not an isolated one. It is typical of many situations in the savanna lands of Africa. A more complete understanding of pastoral issues argues for a re-evaluation of the old orthodoxy, replacing it with new thinking, elements of which have been outlined in this paper. However, a resolution of the problems of pastoral peoples requires concrete political commitment by governments and others to the consequences of this new thinking, most notably in the area of land rights. This may prove more difficult to achieve.

REFERENCES

Abel, N. and Blaikie, P. (1989) Land degradation, stocking rates and conservation policies in the communal rangelands of Botswana and Zimbabwe. *Land Degradation and Rehabilitation* 1:101–23.

Barbier, E. (1989) The economic value of ecosystems: 1. Tropical wetlands. *LEEC Gatekeeper 89–102.* International Institute for Environment and Development, London.

Baxter, P. (1989) *Property, poverty and people: changing rights in property and problems of pastoral development.* Department of Social Anthropology and International Development Centre, University of Manchester.

Bell, R. (1982) The effect of soil nutrient availability on community structure in African ecosystems. In: B. Huntley and B. Walker, (eds) *Ecology of tropical savannas.* Springer Verlag, Berlin.

Bilali, A. (1989) Management of pastures and grazing lands in Tanzania. *Splash* 5(2) SADCC, Maseru.

Birley (1982) Resource management in Sukumaland, Tanzania. *Africa* 52:1–30.

Breman, H. and de Wit, C. (1983) Rangeland productivity and exploitation in the Sahel. *Science* 221:1341–7.

Carter, C., Frank, D. and Loyns, R. (1989) Wheat in African development: The case of Tanzania. Paper presented to Annual Conference of the Canadian Association of African Studies, Carlton University, Ottawa.

Coe, M., Cumming, D. and Phillipson, J. (1976) Biomass and production of large African herbivores in relation to rainfall and primary production. *Oecologia* 22:341–54.

Cossins, N. (1985) The productivity and potential of pastoral systems. *International Livestock Centre for Africa Bulletin* 21:10–15.

Dixon, J. et al. (eds) (1990) *Dryland management: Economic case studies.* Earthscan, London.

Fenger, M., Hignett, V. and Green, A. (1986) Soils of the Basotu and Balangda Lelu areas of northern Tanzania. Agriculture Canada and Canadian International Development Agency, Ottawa.

Frost, P., Menaut, J., Walker, B., Medina, E., Solbrig, O. and Swift, M. (1986) Response of savannas to stress and disturbance: A proposal for a collaborative programme of research. Report of IUBS working group on Decade of the Tropics programme. Biology International, IUBS, Paris.

Hardin, G. (1968) The tragedy of the commons. In: Hardin and Baden (eds) *Managing the commons.* W.H. Freeman, San Francisco.

Herskovits, M. (1926) The cattle complex in East Africa. *American Anthropologist* 28:230–72; 361–80; 494–528; 630–64.

Lane, C. (1988) Beer with the Barabaig. *International Institute for Environment and Development Perspectives* 1:14–15.

Lane, C. (1990) Barabaig natural resource management: Sustainable land use under threat of destruction. UNRISD Discussion Paper, 12. UNRISD, Geneva.

Lane, C. (1991a) Alienation of Barabaig pastureland: Policy implications for pastoral development in Tanzania. D.Phil. thesis, University of Sussex.

Lane, C. (1991b) Wheat at what cost? CIDA and the Tanzania–Canada Wheat Programme. In: Swift, J. and Tomlinson, B. (eds) Conflicts of Interest: Canada and the Third World. Between the Lines, Toronto.

Lane, C. and Pretty, J. (1990) Displaced pastoralists and transferred wheat technology in Tanzania. *Gatekeeper* No. SA20, International Institute for Environment and Development, London.

Lane, C. and Swift, J. (1989) East Africa pastoralism: Common land common problem. Issues Paper No. 8. International Institute for Environment and Development Drylands Programme, London.

MacMillan, R. and Ngoma, A. (1983) Vertisolic soils of the Hanang wheat complex, Tanzania. Characteristics, distribution, use and management. In: Proceedings of 5th meeting of the Eastern African sub-committee for soil correlation and land evaluation. Wad Medani, Sudan. World Resources Reports, 56. FAO, Rome.

119

Makec, J. (1988) *The customary law of the Dinka people of Sudan.* Afroworld, London.

Mbilinyi, R., Mabele, R. and Kyomo, M. (1974) Economic struggle of TANU Government. In: G. Ruhumbika (ed) *Towards Ujamaa: 20 years of TANU leadership.* East Africa Literature Bureau, Nairobi.

McNaughton, S. (1985) Ecology of a grazing ecosystem: Serengeti. *Ecological Monographs* 55:515–25.

McNaughton, S. (1990) Mineral nutrition and seasonal movements of African migratory ungulates. *Nature* 345:613–15.

Mascall, M. and Associates (1986) Report on the evaluation of the benefit/cost report and production cost analysis of the Tanzania Wheat Project. Unpublished Report by Prairie Horizons Limited.

Mustafa, K. (1986) Participatory research and the Pastoralist Question. In: *Tanzania: A critique of the jipemoyo project experience in Bagamoyo District.* Unpublished PhD thesis, University of Dar es Salaam.

National Research Council (1986) *Common property resource management.* Proceedings of a conference, April 1985. National Academy Press, Washington 631pp.

Newman, J. (1970) *Subsistence change among the Sandawe of Tanzania.* National Academy of Sciences, Washington.

Prairie Horizons Ltd. (1986) Final report of the benefit/cost team on the Tanzania Wheat Project submitted to the Natural Resources Branch, CIDA.

Raikes, P. (1981) *Livestock development and policy in East Africa.* SIAS, Uppsala.

de Ridder, N. and Wagenaar, K. (1986) Energy and protein balances in traditional livestock systems and ranching in eastern Botswana. *Agricultural Systems* 20:1–16.

Runge, C. (1986) Common property and collective action in economic development. *World Development* 14 (5):623–35.

Saitoti, ole T. (1986) *The worlds of a Maasai warrior: An autobiography.* Random House, New York.

Schultz, J. (1971) *Agrarlandschaftliche veranderungen in Tanzania.* Welt-Forum, Munich.

Scoones, I. (1990) Livestock populations and the household economy. PhD thesis, University of London.

Scoones, I. (1991) Wetlands in Drylands: The Agroecology of Savanna Systems in Africa. Drylands Programme Paper, IIED, London.

Sinclair, A. and Norton-Griffiths, M. (1979) *Serengeti: Dynamics of an ecosystem.* Chicago University Press, Chicago.

Stone, J. (1982) Project evaluation: A case study of the Canada–Tanzania Wheat Project. University of Guelf, Guelf.

Swift, J., Toulmin, C. and Chatting, S. (1989) Providing services for nomadic peoples: A review of the literature and annotated bibliography. WHO/UNICEF.

Tanzania Government (1982) *The livestock policy of Tanzania.* Ministry of Livestock, Government Printer, Dar es Salaam.

Wilson, Mcl. G. (1952) The Tatoga of Tanganyika. *Tanganyika Notes and Records* No.33.

Young, R. (1983) Canadian development assistance to Tanzania. North South Institute, Ottawa.

CHAPTER 7

BECOMING KENYANS: MAASAI IDENTITY IN A CHANGING CONTEXT

M. Kituyi

SUMMARY

While it is tempting and interesting to analyse traditional land-use systems, the policy-oriented study of pastoral societies and their relationship with savanna ecosystems must start from what is known: how individual production units *currently* confront man and nature in their pursuit of material and social viability. Within East African pastoral society, Maasai pastoralists have been captured by an emergent state and market economy. Change has been inevitable and irreversible. Recognising the strength of competing forces in the national arena and without sacrificing the ecological underpinnings of the traditional systems that have nurtured the Maasai for centuries, tenable policy advice should be geared towards directing change, rather than nostalgically trying to reverse the hand of history.

This calls for inquiry into the direction of change as it occurs and policies that can accommodate and guide this process. This change, to a great extent, is a product of the adjusting behaviour of the pastoralists themselves. Some pastoralists have set up different resource flow systems that draw from within and outside the pastoral sector. Some have become involved in agriculture as well as pastoralism, others in transport. This not only enhances their material wealth, but also their power within the pastoral community. These changes in allocative opportunity and practice are part of a process of integration into a wider polity and economy shaped, mainly, by national influences. For the Maasai, this process is treated as "becoming Kenyans".

THE SETTING

The Kenyan Maasai are a transhumant pastoral people numbering about 250,000. They live in two administrative districts with a total area of 39,500 sq km. One of the districts is known as the Narok and the other is known as the Kajiado. Most of the land resembles a seasonal savanna, although limited plateaux and hills to the west and south-east are much wetter with rains up to double the average 750 mm per year for the whole area. The border to the south is occupied by the Tanzanian Maasai, Kalenjin agro-pastoralists live to the west, and most other directions are occupied by the agricultural Bantu of central Kenya (see Figure 7.1).

But the notion of boundedness implied by these border descriptions does not do justice to the location of the Maasai today. Over the past 2 decades, much of the northern and western regions of the "Maasai districts" have received so many immigrants that the Maasai have become a minority group. One of the most significant factors of Maasai geography is that their land adjoins the southern suburbs of Nairobi. This fact has been of great import to their interaction with the regional market economy, and their vulnerability to a regulative state authority.

INTRODUCTION

The point of departure in this chapter is that Maasai society is undergoing processes of change in some specific and important ways. Many of the changes which have shaped the new Maasai physical and economic space may be irreversible. Observation of those changes and the nature of variations in their intensity suggests the need to develop a comprehensive analysis of the processes involved. This has been done by the author in a study that focusses on features that manifest change within Maasai society. This chapter is broadly based on the findings that arise from that study (Kituyi 1990).

The combination of Maasai transformation and reproduction, integration into a wider system and maintenance of unique cultural institutions and values are all part of the contradictory process of change in becoming Kenyans. To understand what factors underpin changing processes among the Maasai and how those factors are determined, there is a need to understand adaptation and regional interaction between the Maasai and their neighbours in the pre-colonial period before 1920. A dominant factor in shaping local relations was the competitiveness or complementarity of adaptational niches that each ethnic group exploited. The picture of ethnic relations only makes sense if it is treated over time to show how the

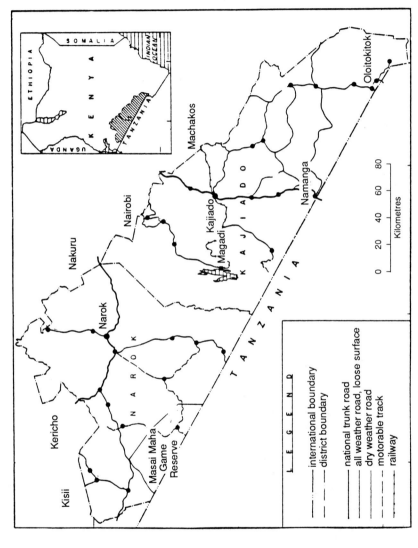

Figure 7.1 The two Maasai districts of Narok and Kajiado

traditional balance of relations has been affected by colonial and market-related changes.

TRADITIONAL RELATIONS AND THE TERRITORIAL FRONTIER

The shaping of Maasai traditional boundaries has involved a complex blend of ecological conditions, the type of adaptation of a given neighbouring ethnic group, military might and negotiation. The most intense relations were with agricultural neighbours, especially the Kikuyu who occupied the tropical forest territory to the north. Although punctuated by raids and civil strife, the dominant character of this relationship was symbiotic. The agricultural Kikuyu produced grains and vegetables to exchange for livestock and livestock products with their herding neighbours. Part of the reason was ecological. The tsetse-infested forests were of no pastoral value to the cattle-keeping Maasai, while the dry seasonal savannas were of limited agricultural value to the farming Kikuyu (Berntsen 1976; Waller 1985).

Interactive processes

The exchange relations were so successful that women from both communities became specialised in the "cross-border" trade, and wider forms of interaction developed. One of the most significant of these relations was the facility by which impoverished and displaced Kikuyu could seek refuge within a Maasai family and gradually get inducted into Maasai society. The counter trend of poor Maasai becoming Kikuyu was also common (Berntsen 1976; Leakey 1977; Routledge and Routledge 1910). To a lesser extent, similar relations pertained for the Maasai with the Chagga and Abagusi cultivators to the south and south-west.

By contrast, relations with pastoral and agro-pastoral neighbours were more dominated by strife. Competition for pasture and livestock raids were more recurrent. In spite of adaptational similarity, the cultural and social exchange between the Maasai and their herding neighbours was much less significant than that with the agricultural Kikuyu, with whom the Maasai shared a common God, rites of passage, and some clan names. Indeed, because of similarity in adaptation, competition characterised the most dominant relationship. This partly explains a tendency to have large buffer lands between the Maasai and such neighbours. These tracts of buffer land are the main domain of the hunter-gathers known as Dorobo.

Whereas the ecology for pastoralism and agriculture is an important factor in determining the frontier between groups, caution needs to be applied in interpreting the ecological frontier between these people. The interaction between pastoral man and his biotic environment is dynamic. Human exchange and ingenuity mediate the limits of ecology in adaptive behaviour and control. It is valid to some extent to see ecology as defined partly by culture and power relations (Galaty and Johnson 1990). It is even more important to see the ecological boundary as modified by technological innovation and population pressure. This point is significant when one examines post-colonial relations between Maasai and their neighbours.

The relevance of this brief historical sketch is obvious when one examines the trajectory of change in regional relations. Important changes for the Maasai have occurred in the image of pre-colonial practice. New dynamics that unequally enhance adaptive capacity to changing conditions have helped upset traditional balance. But the change has often been presented as merely incremental within a pattern internalised by practice. For example, part of the influx of peasant families from central Kenya into Maasai country is due to traditional border flexibility influenced by the southward spread of the Kikuyu who have traditional accommodative relations with the Maasai. This has facilitated techno-economic changes that increased peasant capacity to exploit an ecological niche earlier dominated by transhumant pastoralism.

Analytically, the empowerment of a peasant community through research and innovation (for example, the discovery of drought-resistant crop strains), combined with heightened pressure on rain-fed agricultural land and access to production credit, has changed the nature of competition between peasants and pastoralists. Historically, the Maasai had a symbiotic relationship with their neighbours but with less capacity to adopt new technology, the increased demand for cash and market dependence, has changed the relationship from one of reversal of a symbiosis into one of unequal competition. This unequal outcome is very likely to increase as pastoralists become citizens of a country that places agricultural development high on the political agenda. For many African pastoralists, the reversal of traditional regional relations has been a problem in their becoming integrated into new state entities and facing new challenges (Lane and Pretty 1990).

Ecological costs

The ecological price paid when tropical rangelands are surrendered to encroaching peasant and commercial cultivation has not been given due attention. Although the role played by pastoralism in the reproduction of biodiversity has occasionally been appreciated, the role of alternative land uses needs even greater attention (Kituyi 1985; Western and Dunne 1979).

The conversion of rangelands from protein-producing pastoral lands to cereal growing, while possibly allowing for the sustenance of more people, entails an ecological sacrifice. A complex community of grasses and trees is relinquished for a monoculture of one annual grass with edible seeds. While domesticated animals primarily return the organic matter consumed to the locality in dung and meat, cereal production represents an avenue for stripping rangelands to feed urban mouths. Also cereal cultivation technology has a centrifugal tendency to extend into more marginal areas as soil fertility is used up. In contrast, pastoralism will not use up soils which do not sustain pasture (cf. Darling 1968). There is also an important issue associated with the interdependence of key grazing areas and the extensive grazing systems linked to them. When a key grazing area, such as a valley bottom, is cropped, the potential productivity of the surrounding grazing areas that depend upon it may be reduced drastically (Lane and Scoones, this book).

Changing social linkages

Beyond the transformation of competitive relations with outsiders, the dynamics of private ownership of production resources must also be understood and placed within the context of competition and novel social forms and community relations. The evolution of a market economy has had mixed consequences for the traditional exchange relations that once formed the backbone of Maasai society. On the one hand, the relations are reinforced by new forms of value conversions. On the other hand, they act as ventilators through which adaptation to new exchange and production relations have been pursued (Kituyi 1990).

In discussing the shaping of the Maasai context, another theme is the systematic integration of Maasai territory into Kenya. The physical process of becoming Kenyan by acquiescing to Kenyan government authority is a major process in Maasai history. Gradually, the Kenyan government has established a monopoly over socially legitimate forms of violence and now takes primary responsibility for the maintenance of social order. But, at the same time, becoming Kenyan has opened up new avenues and provided the foundation of a range of acceptable adaptive strategies for the Maasai both within the group and in relation to other Kenyans. Some are taking on a new identity that involves many forms of non-traditional enterprises.

Decision making

Maasai allocate production resources and make choices on a *continuous* basis. For them, the context of resource management entails more than the material resources and the techno-economic factors. It contains the organisational format within which individual decision making occurs.

The organisational behaviour of the Maasai is changing. The institutions that shape the cultural context of the Maasai are themselves meeting varied pressure. This relates to the adaptive patterns that individual management units generate in the context of varying perceived opportunities. Importantly, some of the formal processes that operate at a national level do not permit continuous change. Consequently, new tensions and decision-making frameworks are emerging. Policy reform should only be contemplated once the full implications of these considerations are understood.

New inequities

Emerging disparity in Maasai society is predicated on the articulation between exclusivity of rights to production resources and the resilient constraints deriving from alternative allocations of limited resources (see Boxes 7.1 and 7.2). While emphasising the powerful role of economic factors, it can be argued that part of the explanation of features of regional differentiation is located in the varied decline of ideological decisions made on the basis of the Maasai need to pursue an identity and maintain esteem within the cultural setting. Many resource allocation decisions are becoming more monetarised in a way that is creating new losers and winners. For the losers, desperate economic conditions often lead to unsustainable exploitation of accessible resources (Hjort af Ornas and Salih 1989).

To understand how social institutions and adaptive patterns have diversified, one needs to comprehend and yet transcend a focus on what the government does to the people and examine how people, as economic actors, respond to different policies and pressures. The government, just as the national economy, constitutes a set of variables that greatly influence the shaping of the socio-economic space within which the Maasai devise strategies. Particular emphasis must be given to labour and rights to use land (Kongstad and Monsted1980; Lefebure 1979). People shape their social institutions through the way they appropriate labour and spread it across the resources to which they have access.

The study of Maasai transformation shows an increasing integration of diverse economic activities within the same domestic groups. The intricate ties between varied economic spheres suggest that pastoralism should no longer be viewed as a self-sufficient economy and that it is the fruitfulness

Kwatei is a member of the Purko Olosho (section of the Maasai) living some 10 km west of Narok town. During land registration and the creation of ranches in the 1960s and 1970s members of Kwatei's *enkutoto* locality refused subdivision of their land, hence the ambitious young man had to contend with being a member of a group ranch. In the 1970s, however, using income deriving from the selective sales of livestock and his salary as an elected member of the Narok County Council, Kwatei bought himself a 20-acre farm 30 km west of the group ranch. He married and settled a second wife on this farm where he grows maize.

Kwatei has built a large modern house on the group ranch and fenced off a plot of 7 acres where he grows maize, fruit trees and fodder. He owns over 300 zebu cattle and 10 grade cows. His traditional Zebu cows are mostly herded on the large group ranch and only come to the home in the evening. But the grade cows are kept on his 7-acre enclosure in total exclusion of competitors. Kwatei employs five labourers on a permanent basis. Three of these are involved in the daily herding of animals, and two are agricultural workers tending the crops on his 20-acre farm and on a portion of the group ranch he has started cultivating.

In 1982, he purchased a tractor with credit secured by his 20-acre farm. The tractor is used for a variety of agricultural activities. First, it is used for ploughing and transporting agricultural inputs and produce. Secondly, it is used for transporting water from a stream some 5 km away to water grade animals and to irrigate the fruit trees and fodder crops in the dry season. Thirdly, the tractor transports maize stalks and fodder from the larger farm to the home on the group ranch. The maize residue is a particularly important source of fodder during the dry season, not just for the grade cows but also for all his animals which are fed separately on his enclosed land away from neighbours on the group ranch. Kwatei also buys such stalks from other farmers who have no means of transporting such forage to their own cattle, or who own an insignificant number of cattle. He also transports forage for farmers on the condition that they offer half of it as payment for the transport costs. This "business" of transporting fodder from distant farms has been so successful that Kwatei has never had to sell any animals due to forage scarcity in the dry season.

Box 7.1 Maximising herd and the fodder links. Source: Kituyi (1990), Chapter 7

When Tinkoi retired from the Kajiado County Council in 1965, he had already been in touch with "progressive leaders" who were encouraging the Maasai to acquire individual ranches. Although he wished to have such a ranch, Tinkoi found his own people of Osilale sublocation of Matapato against any subdivision of their common territory. He accepted their position and settled on the collective territory. When the government increased campaigns for ranch subdivision in the early 1970s, however, Tinkoi was one of the six elders locally allowed to claim and register individual ranches.

As a respected elder in the area, Tinkoi has paid particular attention to etiquette in local dealings. One example is his clear effort to show generosity and is revealed in the way he has utilised his individual ranch. Unlike the other individual ranchers in the area, he has left his ranch undeveloped. While he insists on exclusive rights, Tinkoi takes pride in the fact that he never turns down requests from people on the unscheduled Silale Scheme to use his relatively better-watered ranch during the dry season.

Tinkoi sees the ranch as according him two major benefits. First, the ranch has enabled him to own a resource in short supply – dry season pasture. By demonstrating his generosity through allowing access to others when everyone knows he could have done otherwise, he feels he has built the kind of esteem that usually takes many livestock gifts and alliances to build. Secondly, the fast-rising price of land and the fact that he has no sons means that he can get a lot of money through selling the ranch in the future. While others with individual ranches have gone into commodity production and surplus accumulation, Tinkoi serenely drinks his days away with other elders; whilst his modest herd is tended by sons of friends and relatives residing upon or temporarily pasturing on his ranch.

Box 7.2   Capital accumulation: missed opportunity. Source: Kituyi (1990), Chapter 7

to regard pastoralism as an activity that can be separated from other productive practices (Spencer 1977, 1988; Galaty 1981b; Haaland 1990).

Changing skills

While livestock production remains the dominant focus of Maasai labour, non-pastoral activities are becoming an increasingly important part of the

Maasai society. Thus, while livestock remains the focus of valuation and the main representation of accumulated labour in the changing context, proceeds from herds are only a first step in understanding pastoral food production (Hjort 1981).

Analytically, to the extent that adaptive practices emerging within Africa's seasonal savannas are unsustainable, their reversal can not be conceived of as simply removing immigrants who may have demonstrated the short-term viability of the new practices. Increasingly, many formerly alien trades like cropping are being undertaken by Maasai people (see Box 7.1). Thus policy research may need to work out how to compensate for the declining capacity of pastoral production to meet local economic needs, with alternatives that do not deplete the natural resource base. Such alternatives may include some sloughing off of persons from within the pastoral sector (Haaland 1990) and stemming of land expropriation by rich locals, large commercial enterprises, and displaced peasants.

Galaty (1980) has argued that, in studying change and continuity, one way of easing the divide between the two processes is to relate them as figure and ground, with one forming the background for the study of the other, and to emphasise change or continuity depending on the institution, practice or context to be studied. The historical development of the Maasai context has been such that the blend between continuity and change cannot analytically be separated in the way suggested by Galaty. Change and continuity emerge as the diversified field of varied blends from regional, national and local influences.

The fluid nature of social boundaries

In some Marxist anthropological literature, one finds a tendency to arbitrarily erect boundaries between what is seen as the superstructure and the economic base. The trend is then to discern the economic base as an area under transformation with changing property and production relations, but to treat the superstructure as an area dominated by forces of ideological reproduction. Thus, continuity is asserted as a reproduction of the ideological sphere, while change is noted in the economic base.

Bonte (1975, 1981) uses such a perspective in his theoretical conception of continuity and change among East African pastoralists. He sees "religious structures" as dominating the superstructure and being reproduced through various religious and ritual practices. This conception underlies Rigby's (1981) discussion of the failure of Christian churches to evangelise the pastoral Maasai. An extreme variant of this position is to focus on ideology as a force of cultural reproduction to the extent of negating change as a possibility. This is contained in such notions as Freidman's (1976) study of what he calls "systems of total reproduction".

While the need to theorise about the relationship between institutions is obvious, it is not adequate for hastening the creation of institutional boundaries where systems remain greatly undifferentiated. Much less is it the basis of asserting an assumed perpetuity in some areas of social life. Some of this tendency to see culture/ideology as mystification and functionally repetitive is a product of the functionalist background of this line of thought: seeking to identify what it is that maintains the system (Godelier 1972; Ortner 1984).

BECOMING KENYANS

In her study of the Waso Borana pastoralists of northern Kenya, Dahl (1979) treated the historical process of integration characterised by colonial occupation, governmental control, encroaching market and emergence of a new elite as major factors in this process. The alternative view is that the process of becoming Kenyans is seen as a contradictory one containing varied features of integration and exclusion.

Physical, administrative and economic integration is a major component of becoming citizens (see Box 7.3). As the range of investment openings increasingly ties the Maasai adaptation to variables in the national economy, new opportunities and constraints enter their way of life. National integration usually involves a normative proximity that derives from assumptions of shared destiny and values. There is usually the assumed basis of an "imagined community" (Anderson 1983). In a country where there is no developed system of values that bind people to the identity of being Kenyans, however, the variables in "Kenyan-ness" entail changes that need not shape a national consciousness as such. It is in these changes that one may treat the processes discussed in this study as part of the Maasai becoming Kenyans.

The nature of continuity in Maasai culture that has been discussed in this study, and in many others has something to do with the specific policies of British colonialism. It has been argued that one of the consequences of British indirect rule in Africa was that the strategy economised on cultural disruption, which led to the survival of pre-colonial relations in a complementary rather than integrated nationhood (Mazrui and Tidy 1984). The varied survival of pre-colonial avenues of regional exchange, which are important in understanding the changes in Maasai context, relate to this phenomenon.

The articulation between continuity and change in ethnic relations has come to colour the emerging identity of the Maasai. One important area

David comes from the same area as Tinkoi (see Box 7.2) but represents a different kind of entrepreneur. When some individual ranches were carved out of Osilale area in 1973, David, then a young secondary-school drop-out, saw an opportunity to invest in a fruitful enterprise. Despite opposition from older men, he applied for an individual ranch and the government allocated him some 450 ha.

David soon encountered a series of problems. His ranch was the victim of a number of mysterious fires for which he suspected arson. Word was doing the rounds that he was too ambitious and even ignoring the views of the elders. His age-mates soon asked him to demonstrate sensitivity to the wishes of older men, and some friends stopped associating with him. Rather than give up on his dream of ranching, David requested the lands office to take back his ranch and allocate him an alternative one near the Matapato/Purko boundary where more people of varying age had received individual ranches. This was done. He then invested in a bore-hole, fencing and steers. He built a large herd of beef cattle from which he slaughters and sells at the butchery in Bisil. He also transports some cattle to Ngong.

Box 7.3   Social barriers and entrepreneurial pursuit. Source: Kituyi (1990), Chapter 7

of this influence is in the very creation of an ethnic category. Maasai ethnic consciousness has traditionally been associated with their sectional lands. Every domestic group belongs to a local community (*enkutoto*), a number of which constitute the sectional lands (*olosho*) which in turn form the Maasai country. It is only in the changing conditions that some of the boundaries of this consciousness appear to be weakening. This has to do with competition with outsiders and relations within Maasai territories. As the Maasai are exposed to competition with other Kenyans, increased pressure on resources traditionally appropriated by the Maasai leads to a feeling of shared threat from the other people. This translates into closer awareness of being separate and with common purpose. This process has been aided by the increasing interaction within market centres and in economic life in general.

At another level, the Maasai have developed a stake in what goes on at national level. In Kenya, the tendency to identify resource rights with ethnic representation has reinforced the tendency towards ethnic ascription as the basis for claiming a share. As Hjort (1981) notes from the Samburu cousins of the Maasai, this is a context where ethnic ascription becomes "one asset among several to get access to needed or attractive resources, be they subsistence niches in a multi-ethnic context, land rights in a predominantly mono-ethnic area, or some other economic significance". But this integration can also bring new resources, more as imitation of what is seen as desirable elsewhere than as a viable response to local needs.

The accommodation into a new state entity with perceived developmental needs tends to encourage a shopping list of developmental investments that may have no bearing on the varied ecological conditions of the rangelands. Political elites become proponents of, and yet captives to, initiatives that demonstrate power and change. Often these initiatives have ecological consequences that take generations to be adequately comprehended (Ahmed 1989). Hence impact assessment studies of development projects, such as bore-holes, livestock holding centres, townships and irrigation projects, will be critical in relating change and the sustainable use of tropical savannas.

CONCLUSION

By identifying the context where resource access is mediated by ethnic belonging, the Maasai internal divide which is traditionally important in cultural and economic relations is mediated by both new reasons for closing ranks and the fact of every Maasai being a Maasai to the other actors in the emerging multi-ethnic context. Yet some processes also work in the opposite direction.

The competitive nature of resource access has tended to reinforce suspicion between sectional groups. Where disputes on ranch boundaries have escalated to violence, it has been in areas where different sections meet. This increases the awareness of the divide between sections which each have customary rights to live, graze animals and utilise the resources that exist in separate areas. Such an awareness is reinforced by the varied permissiveness to forces of changes which has led to greater entrenchment into a market economy among the northern Maasai, and a common expression by their southern cousins that *iloshon* to the north "are no longer Maasai'.

The argument here is that the levels of identity are shifting to manifest the competition and suspicion noted in pre-colonial relations, yet new bases of ascription emerge as historical circumstances confront the Maasai with other countrymen. These are features notable in other Kenyan societies, such as the ambiguous creation of Abaluhya and Kalenjin as ethnic groups over the past century. In this way, the articulation between reproduction and transformation among the Maasai may be seen as an interaction between features specific to Maasai socio-cultural contexts and forces from the environment externally-regulated. These are processes the Maasai share with other Kenyans but with culture and adaptation-specific consequences on the reproduction of traditional organisational and productive behaviour, and extent of impact on natural resources.

REFERENCES

Ahmed, A.G. (1989) Ecological degradation in the Sahel: The political dimension. In: Hjort af Ornas and Mohammed Salih (eds).

Anderson, B. (1983) *Imagined communities: Reflections on the origins and spread of nationalism.* Verso, London.

Berntsen, J. (1976) The Maasai and their neighbours: Variables of interaction. *African Economic History* 2:1–21.

Bonte, P. (1975) Cattle for god: An attempt at a marxist analysis of the religion of East African herdsmen. *Social Compass* 22 (3–4):381–96.

Bonte, P. (1981) Marxist theory and anthropological analysis: The study of nomadic pastoral societies. In: Kahn and Llobera (eds), pp. 22–56.

Dahl, G. (1979) Suffering grass: Subsistence and ecology among the Waso Borana. *Stockholm Studies in Social Anthropology* 9. Department of Social Anthropology, University of Stockholm.

Dahl, G. and Hjort, A. (eds) (1979) *Pastoral production and society.* L'equipe Ecologie et Anthropologie des Societes Pastorales, Paris.

Darling, F. (1968) Impact of man on the biosphere. Working Document of the UNESCO Conference of Experts on the Scientific Basis for Rational Use and Conservation of the Resources of the Biosphere. UNESCO, Paris.

Friedman, J. (1976) Marxist theory and systems of total reproduction. *Critique of Anthropology* 7.

Galaty, J. (1980) The Maasai group ranch: politics and development in an African pastoral society. In: Salzman, P. (ed.) pp. 157–72.

Galaty, J. (1981a) Introduction: Nomadic pastoralists and social change processes and perspectives. In: Galaty and Salzman, P. (eds) pp. 4–26.

Galaty, J. (1981b) Land and livestock among Kenyan Maasai: Symbolic perspectives on pastoral exchange, change and continuity. In: Galaty and Salzman, P. (eds) pp. 68–88.

Galaty, J. and Salzman, P. (eds) (1981) *Change and development in nomadic and pastoral societies.* E.J. Brill, Leiden.

Galaty, J. and Johnson, D. (1990) Pastoral systems in global perspective. In: Galaty and Johnson (eds) *The world of pastoralism herding systems in comparative perspective.* The Guilford Press, New York.

Godelier, M. (1972) *Rationality and irrationality in economics.* New Left Books, London.

Haaland, G. (1990) Aid and sustainable development in a dual economy. *Forum for Utviklingsstudier* 1:105–25.

Hjort, A. (1981) Ethnic transformation, dependency and change. The Ilgira Samburu of Northern Kenya. In: Galaty and Salzman (eds) pp. 50–67.

Hjort af Ornas and Salih, M. (eds) (1989) *Ecology and politics. Environmental stress and security in Africa.* Scandinavian Institute of African Studies, Uppsala.

Kahn, J. and Llobera, J. (eds) (1981) The Anthropology of Pre-capitalist Societies. Macmillan, London.

Kituyi, M. (1985) The state and the pastoralists: The marginalization of the Kenyan Maasai. DERAP Working paper No. A342. The Christian Michelsens Institute, Bergen.

Kituyi, M. (1990) *Becoming Kenyans: Socio-economic transformation of the pastoral Maasai.* Acts Press, Nairobi.

Kongstad, P. and Monsted, M. (1980) *Family, labour and trade in Western Kenya.* Scandinavian African Institute, Uppsala.

Lane, C. and Pretty, J. (1990) Displaced pastoralists and transferred wheat technology in Tanzania. *International Institute for Environment and Development Gatekeeper Series* No. SA20, London.

Leakey, L.S.B. (1977) *The Southern Kikuyu before 1903. Vol. I.* Academic Press, London.

Lefebure, C. (1979) Introduction: The specificity of nomadic pastoral societies. In: Dahl and Hjort (eds).

Mazrui, A.A. and Tidy, M. (1984) *Nationalism and new states in Africa from about 1935 to the present.* Heinemann, London.

Ortner, S. (1984) Theory in anthropology since the sixties. *Comparative Studies in Society and History* 26:126–66.

Rigby, P. (1981) Pastors and pastoralists: the differential penetration of Christianity among East African pastoralists. *Comparative Studies in Society and History* 23:97–127.

Routledge, M. and Routledge, K. (1910) *With a pre-historic people: The Akikuyu of British East Africa.* Franc and Cass, London. (New Impression in 1968.)

Salzman, P. (ed.) (1980) *When nomads settle: Processes of sedentarization as adaptation and response.* J.F. Bergin, New York.

Spencer, P. (1977) Pastoralism and the ghost of capitalism. *East African Pastoralism* 71–92.

Spencer, P. (1988) *The Maasai of Matapato: A study of rituals of rebellion.* Manchester University Press for the International African Institute, Manchester.

Talle, A. (1988) Women at a loss: Changes in Maasai pastoralism and their effects on gender relations. *Stockholm Studies in Social Anthropology No. 19.*

Western, D. and Dunne, T. (1979) Environmental aspects of settlement site decisions among pastoral Maasai. *Human Ecologist* 7 (1):75–87.

Section 4

Dual common-property and private-property grazing systems

CHAPTER 8

SAVANNA LAND USE: POLICY AND PRACTICE IN ZIMBABWE

M.W. Murphree and D.H.M. Cumming

SUMMARY

Zimbabwe is predominantly wooded savanna. An equable tropical climate resulted in a relatively large immigrant white population, a situation that led to early self-governing colonial status, an entrenched dual land-use system, a dual agricultural economy, and delayed political independence. Against economic and ecological criteria, performance has been mixed. Self-sufficiency in food supply has been achieved but is associated with widespread environmental degradation, particularly in the communal lands.

The inequities in Zimbabwe's dual agricultural and land-use system have made them a dominant consideration in land-use policy, but economic and agricultural production concerns have constrained policies driven solely by equity issues. Macro-economic policies have been biased in favour of commercial and industrial interests, for the benefit of urban populations and at the expense of rural land managers. This analysis emphasises the politico-institutional components of land and resource use and argues that policy must not only be economically efficient and ecologically sustainable but also institutionally acceptable and effective.

It is recommended that policy-making on land use, environmental issues and the economy be more tightly integrated; that policy redress rural/urban macro-economic imbalances; that greater attention be paid to the potential of diverse indigenous natural resource exploitation; and that communal-property regimes of land and resource management be developed in communal and resettlement areas.

INTRODUCTION

Zimbabwe is situated on the high plateau of east and southern Africa and lies wholly within the tropics. There are four main physiographic regions, with the eastern mountains forming a narrow band along the Mozambique border. The rest of the country is characterised by the north-east to south-west watershed – the "highveld" which lies above 1200 m and descends to the Zambezi River in the north and the Limpopo River in the south-east via a series of plateaux, with the middle veld (900–1200 m) giving way to the low veld (below 900 m). The soils are mainly derived from the ancient basement complex underlying the continent and are consequently infertile. Apart from the high rainfall areas of the eastern highlands, the country is predominantly wooded savanna with a mean annual rainfall of between 400 and 1200 mm per annum. Some 65 per cent of the country receives less than 750 mm per annum. The moister, north-eastern sector is able to support commercial farming based on cash and food crops and beef production.

An attractive tropical climate conferred by its altitude (about 66 per cent lies above 900 m) and the promise of high agricultural potential resulted in a relatively large, for an African colony, immigrant white population. This unusual situation led to early self-governing colonial status, an entrenched dual land-use system (see Figures 8.1 and 8.2), a dual agricultural economy and delayed political independence. Within this framework we examine the developing political economy of Zimbabwe and its impact on changing land-use patterns and savanna ecology.

This chapter identifies the principal components that have shaped contemporary savanna land-use policy and practice in Zimbabwe, and provides policy recommendations of national and international relevance. The analysis is loosely structured within the "Three E" framework outlined in Chapter 1. We examine policy components in terms of their economic efficiency and ecological sustainability but have had more difficulty in dealing analytically with the concept of social equitability, since the meaning of this term is largely determined by ideological climate, political stances or the value preferences of analysts. For Zimbabwe, with its legacy of a racially defined dual land-use system and agricultural policy, social equity is clearly of high political salience and structures political objectives in a variety of dimensions. For the purposes of this paper, however, our use of the concept places more emphasis on its implication of a negotiated consensus, of collective compliance based on perceptions of a balance between individual or sectoral interest and common good. This consensus and compliance with regulatory mechanisms and structures is the central link between policy and practice and determines the long-term efficiency and validity of policy. It is not sufficient for land-use policy to be

140

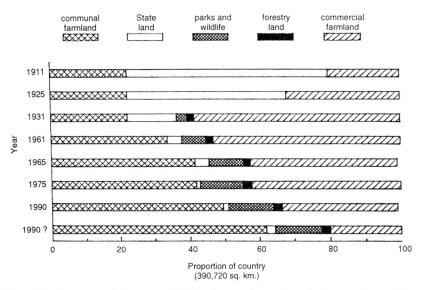

Figure 8.1 Summary of changes, and likely changes, in statutory land apportionment in Zimbabwe: 1911 to late 1990s. Resettlement land is included within communal land and small scale commercial farm land within commercial farm land. (Adapted from Kay 1970; and Cumming 1991a)

economically efficient and ecologically sustainable; it must also be institutionally acceptable and effective. This perspective leads our analysis to place considerable emphasis on the politico-institutional components in land and resource use.

EVOLUTION OF SAVANNA LAND USE IN ZIMBABWE: 1890–1980

Agro-pastoral systems began in Zimbabwe about 2000 years ago when domestic livestock first reached southern Africa. The period from AD 700 to 1700 saw the development of Mazimbabwe (towns and fortifications built in stone) and associated crop and livestock production in an arc extending along the eastern watershed of the country from the north-east to the south-west. These settlements were associated with granite areas and the lighter, more readily tilled, soils. They were also largely beyond the reaches of tsetse fly (*Glossina* sp.). By the fifteenth century the power of Great Zimbabwe had declined and was replaced by the Torwa states and cities of Matabeleland. At the time of colonisation in the late nineteenth century the Matabele had established their power over much of what is

141

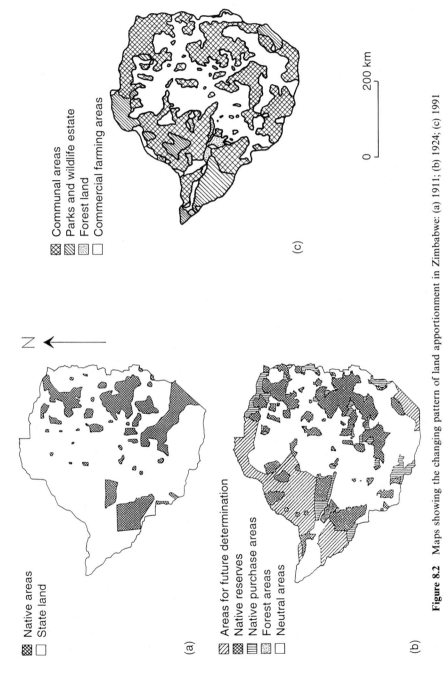

Figure 8.2 Maps showing the changing pattern of land apportionment in Zimbabwe: (a) 1911; (b) 1924; (c) 1991

Communal areas
Parks and wildlife estate
Forest land
Commercial farming areas

(c)

200 km

0

N

Native areas
State land

(a)

Areas for future determination
Native reserves
Native purchase areas
Forest areas
Neutral areas

(b)

now Zimbabwe and had developed a predominantly cattle-based agro-pastoral economy. Livestock was traded as far away as Kimberley. There is no evidence of truly pastoral systems ever developing in Zimbabwe (Beach 1977).

The early White settlers who moved into the country on the pretext of mining concessions soon took to cattle ranching and buying stock from local people. After the Matabele rebellions of 1893 and 1895 stock holdings were seized as the spoils of war and the commercial cattle ranching industry began. This was set back in 1896 by the great rinderpest pandemic that reduced livestock and most game populations to very low levels.

The competitive interaction between indigenous livestock production and settler commercial ranching was thus established at the onset of colonisation. Competition for land for cropping was to follow later. The socio-political developments during this period and their environmental consequences are outlined below.

SOCIO-POLITICAL DEVELOPMENTS

The area between the Zambezi and the Limpopo now known as Zimbabwe was, at white occupation in 1890, inhabited by an indigenous population of 600,000–700,000, with 80 per cent being Shona-speaking and 15 per cent Ndebele-speaking. The Ndebele in the south-west were under centralised rule, the Shona in the centre and east were organised under fragmented polities or chiefdoms, typically involving anywhere from 10,000 to 30,000 inhabitants. Population concentrations occurred in areas where soils and rainfall patterns conducive to dryland agriculture prevailed. These were, however, not the only conditions for settlement. The need for domestic and livestock water supplies, for defensible locations in times of political strife and major demographic movements during the eighteenth and nineteenth centuries were also significant factors which created imbalances in population distribution (Beach 1977).

The base for both the Shona and Ndebele economies was agriculture, the staple crops being millet, sorghum and to a lesser extent maize, all under hoe cultivation. This base was, however, vulnerable to periods of drought, to plagues of locusts or birds or even to floods and constrained by inadequate means of grain storage. Hunting and gathering were thus important subsidiary activities. Amongst the Shona iron smelting, gold mining and a degree of internal and external trade mitigated the uncertainties of agricultural production (Beach 1977). The main insurance against such uncertainties was domestic livestock, particularly cattle. Cattle were largely restricted to the tsetse-free areas of the Matabeleland and Mashonaland highveld, but, particularly in Matabeleland, they played a significant role in domestic economies. Palmer estimates that in 1890 the Ndebele held a herd of a quarter of a million head (Palmer 1977). This was equivalent to 50 per cent of the total livestock population in Zimbabwe at that time (Beach 1977).

A human population of 650,000 in 1890 suggests an overall population density of 1.67 persons per sq km. Coupled with the population clustering mentioned above, this clearly implies that large areas of the country were "unoccupied" in terms of agricultural or grazing activities. It is important to recognise, however, that none, or very little, of this land was regarded as "open access" (*res nullius*) land. Most land was under some form of putative claim, the boundaries of the various polities being contiguous and defined, supported by group mythologies legitimating occupancy and proprietorship. These polities or their subdivisions thus constituted bounded areas of common property under traditional and collective regimes of common-property management (*res communis*). Within these areas arable land was allocated by traditional authorities to individual household heads, such land acquiring a quasi-privatised status, providing permanency of tenure (Cheater 1990). Beyond such lands the use of the grazing commonage for livestock was regulated by communal sanctions and demarcated where necessary to distinguish between the rights of a community and those of its neighbours. This common-property regime was further extended to "unoccupied" land beyond the grazing commonage, to land containing resources used in hunting and gathering and land containing "key resources" useful in times of stress (Scoones 1989). Sometimes, access to these unoccupied lands was the subject of reciprocity with other units of proprietorship.

The regime of land and resource utilisation was informed by an ethno-ecological knowledge system evolved in the micro-environments of the region. Compliance with the regime was effected largely through the dynamics of social conformity, formalised by the weight of religious sanctions. The founding ancestral spirits were regarded as the "owners of the land", controlling the incidence of rain, the fertility of the soil and fecundity of natural resource production. These ancestors, through the spirit mediums (whose task it was to coalesce and articulate collective interest), laid down regulations governing the use of natural resources – the planting of crops, the take-off of fish and wildlife, the cutting of trees and use of fire. Breaches in these regulations resulted in the withdrawal of the benevolence of the land and its resources; we have here, in Rappaport's phrase, "a ritually directed ecosystem" (Rappaport 1969). Beyond this, the religious system closely linked public immorality with ecological irregularity, usually demonstrated during a drought. Schoffeleers (1978) characterises this as a view that the management of nature depends on the correct management and control of society. He goes on to state:

144

This is a profound intuition, and it is also one which is at the heart of ecological thinking in African societies. It is a concept which the industrialized world has largely lost but which it has to restore to its rightful place if it desires a lasting solution to its ecological woes. Keeping or restoring the balance of the ecosystem is not only a matter of technology but also of an ordered social life (Schoffeleers 1978).

What emerges from the above is a generalised picture of pre-colonial savanna land use where agricultural production was paramount, but supplemented by other forms of natural resource exploitation. Management was through common-property regimes, clearly bounded and with explicit rules of inclusion and exclusion, rights and obligations. Given the circumstances of the time, these regimes were economically viable, ecologically sustainable and organisationally efficient, compliance being internally generated rather than externally imposed.

These circumstances changed with the coming of White rule under the British South Africa Company in 1890, although the changes were not immediately apparent. Indeed the period 1890–1908 has been dubbed the "era of peasant prosperity" for Shona farmers as they turned their agricultural skills to the production of surplus crops for sale to growing populations in the towns and on the mines. By 1908, however, the British South Africa Company had abandoned its hopes for "a Second Rand"[1] in Rhodesia and set a policy of diversifying the economy and encouraging White agriculture. As it evolved over the following 25 years this policy had a number of dimensions. The first was the appropriation of the most fertile and advantageously located land for White use, both as a factor of production and an object of speculation (Arrighi 1973). Concomitantly, a system of communal Black lands was established (successively called "native reserves," "tribal trust lands" and "communal lands" in the country's land terminology), rationalised on the grounds that it protected Black communal interests against the avariciousness of White land speculators. The efflorescence of this policy is found in the Land Apportionment Act of 1931, which made explicit an allocation of 198,539 sq km to 50,000 Whites and 117,602 sq km to 1,080,000 Blacks, the balance of 74,859 sq km being reserved for national parks, forestry land and state land (see Table 8.1). The indigenous population had thus lost proprietary rights to more than two-thirds of the land they held in 1890, most of this being the best agricultural land located near urban markets and service centres (see Figure 8.2).

Loss of land was not the only mechanism that barred Blacks from

---

[1] The Witwatersrand formed the gold mining centre of southern Africa on which the city of Johannesburg developed.

Table 8.1 Land allocation (sq km) in relation to natural region and land tenure in Zimbabwe 1990

Region	Commercial farm land				Communal farm land		Parks and forests		Total	
	Large scale		Small scale							
	Area	*%*	*Area*	*%*	*Area*	*%*	*Area*	*%*	*Area*	*%*
I	4402	62.6	73	1.0	1283	18.2	1276	18.1	7034	100.0
II	43,245	73.8	2521	4.3	12,551	21.4	291	0.5	58,608	100.0
III	32,406	44.5	5361	7.4	28,147	38.6	6963	9.6	72,877	100.0
IV	40,258	27.2	5320	3.5	73,073	49.4	29,269	19.8	147,920	100.0
V	36,484	34.9	976	0.9	47,740	45.7	19,211	18.4	104,411	100.0
Total	156,795	40.1	14,251	3.6	162,794	41.7	57,010	14.6	390,850	100.0

commercial agriculture. Policy ensured that White agriculture was heavily subsidised through loan facilities, pricing policies and the provision of extension and infrastructural services. Black agriculture, up to the 1950s, enjoyed none of these advantages and was further restricted through limitations placed on the marketing of agricultural produce from the communal areas. Black farmers were squeezed out of commercial agricultural production and forced to become wage labourers in industry, mining and on White farms – which was precisely what the racially structured economy demanded. The communal lands became, in effect, a residential and subsistence base for the predominately Black wage labour-force of the economy. This was particularly true of the post-employment needs of wage labourers, enabling wages to be set at levels that did not take this requirement into consideration. As Harris commented in 1974, "... the rural areas act as a social security system for industrial workers, supporting the entire family in the eventuality of the head of household losing his position, and in particular supporting the family in retirement" (Harris 1974).

Within the communal lands, the official perspective was that land and resource use would continue to be governed by common-property regimes of management through traditional authorities. In effect, however, the conditions that made such regimes effective in the past had been seriously eroded. Much of the land base had been removed, restricting the flexibility and adaptability of usage which ecological conditions required. Proprietorship over a range of remaining natural resources had also been expropriated, notably wildlife, certain timber resources and minerals. Perceptions of security of tenure, both individual and collective, had been shattered by large-scale translocations and by imposed land-use planning within communal lands by government agents. Finally, it became patently clear to local communities that they, and their traditional leadership, no longer had the autonomy required for effective self-management. The effect was that in most communal lands the mechanisms of collective conformity were curtailed and elements of an "open access" perspective developed, with individual entrepreneurship invading the commons as a collective sense of proprietorship was lost.

Land alienated to Whites was privatized in large agricultural holdings, typically c.1000 to 1500 ha in high rainfall areas or in much larger units (e.g. about 10,000 ha) in areas considered suitable only for ranching. In contrast to the property-rights regime in communal lands, white agricultural land was under a private-property regime. Using a Hardin-type analysis (Hardin 1968), such a regime would promote efficiency of management and also ensure long-term sustainability. The judgement on this thesis from Zimbabwean history must be a mixed one. White agriculture, based largely on the production of maize, tobacco and beef, has undergone fluctuating fortunes and, at several stages in its history, has had to be rescued by state intervention. At the same time, it must be noted that current large-scale owner-operator agricultural units in the country show a measure of investment, innovation and efficiency which the state is reluctant to abandon. Judgement on the second count of ecological sustainability must also be mixed. During the first 4 decades of this century, privatised proprietorship and a relative abundance of natural resources did not necessarily lead to a long-term, sustainable exploitation regime. Indeed the very abundance of resources and their apparent inexhaustability led to a laissez-faire, "mining the soil" approach characterised by the stripping of tree cover, overgrazing and tillage methods that encouraged erosion. This approach was colourfully encapsulated by a member of the Southern Rhodesian Legislative Assembly in 1938 when he said that, "There is a typical Rhodesian attitude towards these things (conservation) which may be expressed by the phrase ... "after me, the desert" (Phimister 1986).

Conservationist concerns over this state of affairs in official circles resulted in the Natural Resources Act of 1941, which provided for a board with powers to compel "farmers and landowners to construct and maintain soil conservation works, protect the sources and courses of streams, control water, regulate depasturing of stock ... adopt approved methods of cultivation, or prohibit or restrict cultivation on any particular piece of land" (Phimister 1986). Under the Act conservation concerns were administered in White farming areas by Intensive Conservation Area committees (ICAs), typically involving 80–90 farms. ICAs were set up when a majority of farmers in the area concerned agreed, thereby becoming eligible for enhanced state subsidies. ICAs were particularly effective in implementing what Phimister terms "mechanical conservation" and by 1950 it was estimated that 91 per cent of all cultivated land on White farms was protected by contour ridging. By contrast "biological conservation," that is the integration of livestock farming and field husbandry, grazing control and ecologically balanced exploitation lagged behind. The reasons for this delay were twofold. First, biological conservation techniques were difficult to enforce and, second the policy conflicted with larger national production concerns (Phimister 1986). It is nevertheless clear that the ICAs have played a dynamic and positive role in contributing to the current economic prosperity and environmental stability of the commercial farming areas. The effectiveness of this role, however, cannot be attributed solely to the existence of private tenure regimes. The relative abundance of available resources; and access to state subsidies (direct or indirect), must also be taken into account. Importantly, ICA committees represented collective regulatory units of manageable size, through which national conservation interests could be modulated through local organisation.

The Natural Resources Act also applied to communal lands, but here implementation was not through representative local organisations analogous to the ICAs but rather primarily by government agricultural extension services. As noted above, government had been reluctant to provide such services earlier in the century, but by the 1950s an agricultural extension agency for the communal lands had been developed that, by comparative standards, was both active and extensive. The attempted mode of implementation of the Act was authoritarian; its motivation stemmed both from conservationist concerns and also from the imperative to maintain food supplies. This technicist and authoritarian approach, coupled with the disabilities of the land and resource management regime in the communal lands described earlier, did little to alleviate the situation. Communal land inhabitants were not in a position to respond as their land base was inadequate and their capacity to implement collective and localised controls had been emasculated.

Some recognition of this dilemma was found in official circles. More land was transferred from state to communal land categories and a longer-term solution was seen in stabilising the communal land population by setting a limit on individual entitlements, the surplus being absorbed in an urbanised population supported by the country's growing secondary industries. This approach found its legal expression in the Native Land Husbandry Act of 1951 which sought to register and limit grants of individual farming and grazing rights, and to cluster settlement to facilitate the provision of services. In Yudelman's view, the Act proposed "to replace the tribal–communistic system of allocating land according to need with a hybrid tribal–capitalistic system of individual holdings and communal grazing" (Yudelman 1964). The farming and grazing rights were not, however, secure. They could be lost if the provisions of the Natural Resources Act were not followed; thus the conservationist component in the Act became the vehicle for perceptions of continued insecurity of tenure. Furthermore, the Act in its planning and attempted implementation by government agents was seen as an exogenous imposition. It had no roots in local consensus and, therefore, lacked a localised institutional base for implementation. It was never implemented at the planned pace and was formally suspended in 1962. Its principal legacy over the following decade was to act as a rallying point for nationalist politics, which saw it as an example of land-use planning and conservationist concern prostituted to White politico-economic ends.

ECOLOGICAL COMPONENTS AND CONSEQUENCES OF LAND-USE POLICY

Major constraints to land use in Zimbabwe are summarised in the description of agro-ecological regions developed by Vincent and Thomas (1961). This classification also provides a convenient, although somewhat overgeneralised, basis on which to examine land capability and use in Zimbabwe. For the last three decades the system, updated by Zimbabwe's agricultural extension service (Agritex), has formed the major basis for land-use planning and analysis. Five agro-climatological regions (see Figure 8.3) are recognised as follows:

Region I   In the *Eastern Highlands*, covering less than 2 per cent of Zimbabwe. Rainfall above 1000 mm. High altitude and low temperatures enable afforestation and intensive diversified agriculture including tea, coffee, deciduous fruits and intensive livestock production.

Region II   The *northeastern-highveld* covering some 16 per cent of the country. Reliable rainfall of 750–1000 mm between November and March; suitable for intensive cropping and livestock production.

Region III Mainly in the *midlands* and covering 18 per cent of the country. Rainfall between 500–750 mm but subject to mid-season dry spells and high temperatures; suitable for drought-resistant crops and livestock. Semi-intensive farming.

Region IV   *Low-lying* areas in the north and south of the country and covering 37 per cent of Zimbabwe. Rainfall between 450–650 mm. Subject to periodic seasonal droughts and severe dry spells during the rainy season. Generally unsuitable for dryland cropping and suited to livestock production.

Region V   *Lowland areas* generally below 900 m and covering 27 per cent of the country. Erratic rainfall usually below 650 mm. Suited to extensive livestock production or game ranching.

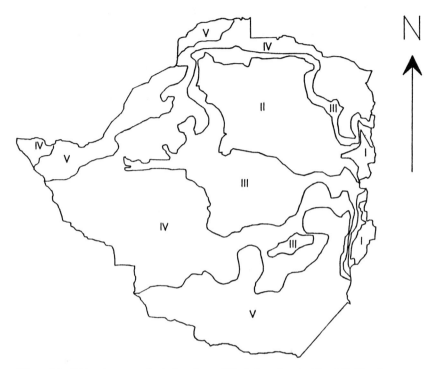

Figure 8.3   Natural agro-ecological regions of Zimbabwe (see text for definitions)

Rainfall emerges as the greatest physical constraint to agricultural production in Zimbabwe and arrives in a single rainy season (November to March) with about 65 per cent of the country receiving less than 750 mm per annum. The combined constraints of rainfall and land capability (soil type and slope) reduce the area of arable land available for intensive dryland cropping to about 7 per cent of the country. The overall area of arable land is estimated to be about 22 per cent (Republic of Zimbabwe 1982). The distribution of arable land by land tenure is summarised in Table 8.2. However, the constraints of rainfall reliability and length of growing season (see Table 8.3) in low-lying and lowland Regions IV and V preclude the productive use of much available arable land.

Zimbabwe lies between altitudes of approximately 300 m and 2800 m with nearly two-thirds of the country above 1000 m. Mean annual temperatures are about 18–19°C. This means that temperature is not normally a limiting factor in plant growth. Frost occurs in low-lying areas, particularly in the arid south-west and can have major effects on woody vegetation, grazing and on winter-irrigated crops in the higher and moister areas of Zimbabwe.

Table 8.2 Distribution of potential arable land and areas (in sq km, percentages in parentheses) cultivated in 1980

	Soil				
	Arable	*Cultivated*	*Crops*	*Fallow*	*Irrigated*
Commercial farmland					
Large scale area	48,000 (55.8)	10,200 (29.1)	6200 (24.5)	4000 (41.0)	1517
Small scale area	5000 (5.8)	900 (2.6)	700 (2.8)	200 (2.1)	–
Communal farmland					
area	33,000 (38.4)	24,000 (68.4)	18,450 (72.8)	5550 (56.9)	–
Total	86,000	35,100	25,350	9,750	1,517
Percentage of					
Zimbabwe	22.0	9.0	6.5	2.5	0.4

Table 8.3 Growing season for dryland crops in natural regions III, IV and V in Zimbabwe. Source: Hussein (1987)

Seasonal parameter	Region III	Region IV	Region V
Mean median start	10 Nov	18 Nov	3 Dec
Mean median end	31 Mar	28 Mar	24 Mar
Mean median length (days)	131	121	96
Mean % occurrence of drought season	18	16	18
Mean % occurrence of intermediate season	5	15	30
Mean % occurrence of no rainy season	0	1	5

Almost all soils in the country are deficient in nitrogen, phosphorus and sulphur, and short (2–3-year) resting fallows covered by weeds and grasses are insufficient to restore fertility to cultivated land in the way that longer (15 year) traditional bush fallows did in the past (Grant 1981, 1987). Fertile irrigable soils are limited in distribution, with the most extensive areas occurring in the arid south-east lowveld.

Apart from soil fertility and nutrients, major constraints to sustainable cropping and livestock raising are imposed by combinations of terrain, soil type and rainfall patterns through their influence on soil erosion. Stocking and Elwell (1973) derived an erosivity index and applied it to the country as a whole. Their map (see Figure 8.4) indicates that some 30–40 per cent of the country, and nearly all of the Zambezi valley, is subject to a high risk of erosion.

The changing patterns of statutory land allocation and use from 1890 to 1980 (see Figure 8.1 and Figure 8.2) when viewed in the context of land

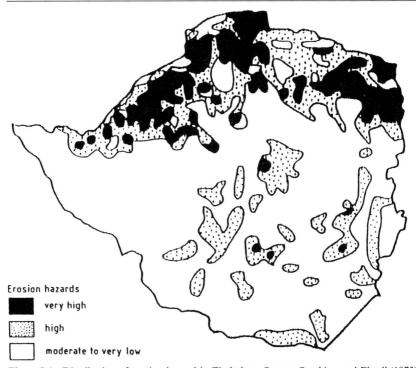

Erosion hazards

■ very high

▦ high

□ moderate to very low

Figure 8.4 Distribution of erosion hazard in Zimbabwe. Source: Stocking and Elwell (1973)

capability reveal that much of the prime agricultural land is not being used and much of the land under cultivation is not capable of sustained cropping.

The marked disparity between the quality of land in commercial and communal farming areas (see Table 8.1, Figure 8.5) is exacerbated by the great difference in population density in the two categories of tenure. In large-scale commercial farming areas, the human population density of landowners and resident labour in 1982 was about 7 per sq km while in the communal lands it was about 26 per sq km. In the small-scale commercial farming areas it was 12 per sq km (Whitlow 1988).

The broad ecological implications of land allocation and agricultural policies described in the previous section are now reflected in corresponding patterns of deforestation and erosion in Zimbabwe (Whitlow 1980, 1988). The differences between commercial and communal farming areas in levels of erosion (see Figure 8.6) are also reflected in levels of deforestation and overgrazing. The often stark contrasts between communal and commercial farm lands reflect not only the result of differing policies but also a long history of differing technical approaches to the question of land use and practice. In large-scale commercial areas the dominant thrust was towards

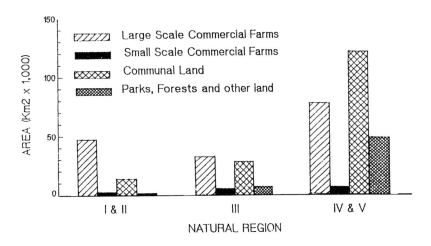

Figure 8.5 Land allocated to commercial and communal farming and other uses in relation to natural region in Zimbabwe (1980)

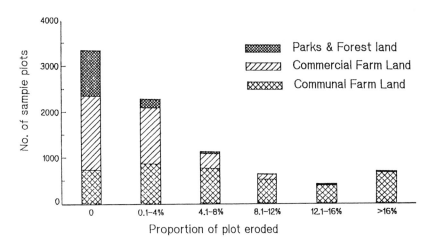

Figure 8.6 The incidence of different levels of erosion in communal land, commercial farm land, and other land (national parks, wildlife land and forest land). Plot sizes were approximately 45 sq km with erosion assessed on an aerial photograph within a grid covering 1000 cells each equivalent to 1 ha. (Data calculated from Whitlow 1988)

conventional, specialised high-input agricultural crop or livestock production. The integration of diverse production systems and the introduction of farming systems research were delayed until the 1980s and environmental considerations beyond soil conservation were generally neglected. The technical approach towards the communal lands focused on soil conservation and on attempting to increase productivity with limited local inputs (manure) on progressively scarce land resources and minimal infrastructural development.

Communal farm land

Four main technical phases of pre-independence development in the communal lands, suggested by Hughes (1974), were:

(1) Stabilisation, which covered the period 1900 to 1925 and was concerned to stabilise what were then termed native land holdings and to reduce shifting cultivation and movement of African settlements. It was during this period that the first native reserves were declared.

(2) Centralisation, which covered the period 1930 to 1950 during which there was an attempt to bring households together in villages of around 15 families with surveyed and demarcated 1-acre plots, in lines, for the households and similarly surveyed and pegged fields for cropping and kraal sites. Grazing areas were designated and separated from those of neighbouring villages.

(3) Individualisation (of tenure) was attempted in the early 1950s through the Land Husbandry Act of 1951. This built on and extended the centralisation programme but incorporated measures that alienated communal land to individual households. Although considered technically sound at the time, the social consequences were not acceptable. The programme failed and was abandoned towards the end of the decade.

(4) Localisation (of decisions regarding land use) was an attempt during the 1960s and early 1970s to devolve responsibility for land-use decisions to local levels through the traditional leadership of chiefs and village headmen. Technical support was to be provided by government extension and community development specialists. The programme made little headway and dissolved with the onset of the independence war.

Ecological impacts of successive interventions in this sector are largely a result of human population growth and lack of inputs, infrastructure and markets which served to maintain, if not impose, subsistence agricultural practices and a subsistence economy on the communal lands. The level of erosion, deforestation and general land degradation in the communal lands is closely related to population density (Whitlow 1988).

Commercial farm land

The following major technical stages in agricultural policy are apparent:

(1) The initial, laissez-faire, phase (1895–1930) was characterised by the largely uncontrolled development of cattle and tobacco farming. A major ecological impact during this period was deforestation of the miombo woodlands to provide fuel for curing tobacco. Large areas of the Mashonaland highveld and the Midlands were affected.

(2) The introduction of soil conservation programmes, market controls and pricing policies characterised the period from 1930–1945. This period, through the implementation of the Natural Resources Act, led to improved management of soils on commercial farms and addressed problems of overgrazing.

(3) The post-war settlement of "Crown Land" and increased production of tobacco, maize and beef during the period 1945 to 1965 re-introduced the large-scale clearance of miombo woodlands to provide fuel for flue-curing tobacco. Systems of rotation of fields and fallow also led to an increase in areas cleared for cultivation. Declining viability of tobacco under marginal conditions, such as in the midland Region III, led to the abandonment of fields and the regrowth of dense stands of mixed miombo woodlands and *Terminalia* scrub.

(4) Diversification of production (1965–1975) was stimulated by international sanctions and the need for self-sufficiency in food production and to circumvent the dependence of the agricultural industry on tobacco exports. This period was also characterised by the development of game ranching and wildlife utilisation on commercial ranches (Child 1988; Cumming 1989, 1991a).

Other important technical interventions on behalf of commercial agriculture, included veterinary control measures and impoundments for hydroelectric power and irrigation. Both forms of intervention had major ecological impacts and were not confined to a particular period in history.

Veterinary control measures were primarily aimed at protecting the commercial livestock industry. As early as 1919 game elimination programmes were started in an effort to stem the spread of tsetse fly that returned with the recovery of game populations after the rinderpest epidemic of 1896. Compulsory dipping of cattle was introduced in 1922 and contributed to the expansion of livestock. During the 1950s control through game elimination was replaced with bush clearing programmes that had long-lasting effects on riverine vegetation where they were applied. In 1963 game elimination was re-introduced as a method of control but focussed on selected species (elephant, buffalo, kudu, bushbuck, warthog and bushpig). They were systematically eliminated from fenced corridors 5–20 km wide, extending along the tsetse front in the north and south of the country. The game elimination programme was increasingly supplemented by dichlorodiphenyltrichloroethane (DDT) ground spraying. During the 1980s hunting was completely replaced by chemical control methods employing ground spraying, aerial spraying and the use of odour-baited traps. Tsetse fly has been eliminated from some 100,000 sq km of land since 1963. The tsetse control programme reduced the numbers and distribution of indigenous large mammals in Zimbabwe. The system of access roads needed to service fences and spraying operations greatly facilitated uncontrolled settlement in otherwise inaccessible and remote areas.

The control of foot and mouth disease of cattle was accompanied by control of wild mammals by hunting and cordon fences. Some 6000 buffalo were eradicated from commercial and communal farming areas in the south-east lowveld during the 1970s – a policy which later had markedly adverse effects on the developing game-ranching industry in the region (Child 1988). Game-proof cordon fences extended from the Nata River in the south-west to the Zambesi Valley in the north-east and encircled the Gonarezhou National Park in the south-east of Zimbabwe. Secondary cordon fences, used to control cattle movement, extended over more than 2000 km.

The development of major hydroelectric power and irrigation schemes began during the late 1950s and the 1960s. Large-scale irrigation projects were introduced into the arid regions of the south-east of the country and the Kariba dam was built to provide hydroelectric power. Both projects involved the development of large impoundments. In the case of the Kariba dam, this entailed flooding large areas of alluvial habitat along the Zambezi River and the displacement of Batonga peoples from the river to the hinterland. The subsequent influx of people and the expansion of subsistence agriculture in the wake of tsetse fly control in the region provided an important example of land-use dynamics, following large-scale interventions in development and disease control. The major result of these developments in the Sebungwe[2] over the last 30 years has been the settlement of large areas of land with

---

[2]The area covered by Binga, Gokwe and Kariba districts in the north-west of Zimbabwe

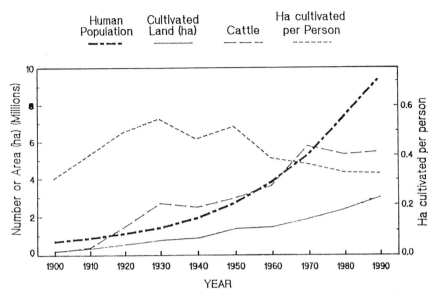

Figure 8.7 Human population growth and accompanying growth in livestock and cultivated land in Zimbabwe since 1900. The per capita area (ha per person) under cultivation is also shown

the highest erosion risk (Figure 8.4) by subsistence farmers and the deforestation of protected areas by overpopulations of elephants (Cumming 1980). The two main forms of land use in the region are subsistence agriculture and parks and wildlife land – the juxtaposition of which has been instrumental in generating the Communal Areas Management Programme for Indigenous Resources (CAMPFIRE). This programme seeks to devolve management responsibility for, and benefits from, natural resource management, particularly wildlife to local communities (Martin 1986).

By 1980 the overriding feature of land use in Zimbabwe, and therefore of the environment and its conservation, was the apportionment of land and the duality of the agricultural sector. The inequity of land distribution in relation to land capability and human population density placed the land issue and rural development at the centre of the government's development programme. The magnitude of the social and environmental problems is well illustrated by the distribution of commercial and communal farm land in relation to agricultural production potential (see Table 8.1 and Figure 8.5). The land apportionment policy started at a time when human populations were very low, and its impact has since been greatly exacerbated by human population growth and the necessary accompanying growth in livestock and cultivation (see Figure 8.7).

LAND-USE POLICY SINCE 1980

The major objectives of government agricultural policy as enunciated in the Transitional National Development Plan (1982) were:

(1) An acceptable and fair distribution of land ownership and use;

(2) A greater degree of economic security and welfare for the rural population;

(3) An increase in both land and labour productivity in all agricultural systems;

(4) A substantial increase in employment to engage a rapidly growing labour force;

(5) Achievement and maintenance of food self-sufficiency and regional security;

(6) Extension of the role of agriculture as a major foreign exchange earner and source of inputs to industry;

(7) Integration of the commercial and peasant agricultural sectors into a national agricultural system;

(8) Conservation of land and environment for future generations;

(9) Promotion of local markets and integration of trade; and

(10) Development of human resources in rural areas to the full.

These objectives were to be translated into the following policies and programmes:

(1) Land resettlement;

(2) Reform and expansion of structures of complementary services including agricultural credit, marketing, research and extension;

(3) Establishment of a number of production systems, including;

    (a) communal farming co-operatives

    (b) private/family and co-operative farms of a variety of sizes

    (c) state farms;

(4) Pursuance of appropriate agricultural pricing policies to achieve the objectives of food self-sufficiency and the extension of the role of agriculture as a significant foreign currency earner;

(5) A closer alignment of land-use patterns and land capability;

(6)   Promotion of research in appropriate technology;

(7)   Deployment of various means at government's disposal, including the land utilisation tax, to ensure that unused and underutilised land and surface and groundwater are efficiently used;

(8)   Development of water resources, elimination of tsetse flies, improvement of conditions of health in all areas, and promotion of research into suitable crops for arid areas; and

(9)   Encouragement and promotion of the establishment of small and medium scale agro-industries (Republic of Zimbabwe 1982).

The central and overriding issue, however, was land allocation and redistribution.

Land distribution and resettlement

On its accession to power in 1980 the ZANU-PF Government had as a central item on its agenda the issue of land redistribution. As the ZANU-PF Leader, Mr. Mugabe, later commented, "the land question was at the centre of the factors that propelled us to launch our war of national liberation" (*The Herald*, 20.12.89). Action on this issue was therefore a political imperative in terms of the government's promises to its constituency, as well as a concern in terms of social justice and longer-term politico-economic stability. At the same time, there were constraints on any whole-scale and rapid transfer of land from the large-scale commercial farming sector to communal land farmers. These constraints included the provisions of the Lancaster House Agreement which stipulated that land could be acquired by government for resettlement from commercial farmers only on a "willing seller, willing buyer" basis during the 10-year life span of the agreement. More importantly, the new government was concerned to ensure that the agricultural productivity of the country was maintained and that the country's vital food supplies were protected. Palmer notes that in 1980 the White commercial farmers were producing some 90 per cent of the country's marketed food requirements and comments that, "at this precise and important moment in time, they seemed crucial to Zimbabwe's economic survival" (Palmer 1990). A third constraint existed in the new government's ideological, bureaucratic and conservationist concerns. Commercial farming land was not to be made available for resettlement simply to replicate on a larger scale existing communal land conditions. Resettlement aimed to promote collectivised modes of agricultural production, planned and implemented in line with sound ecological, agricultural and technical requirements and managed by state bureaucratic and technical agencies.

Table 8.4 Land classification by use: 1989 and late 1990s

	1989		Late 1990s	
Land use	ha × 1000	%	ha × 1000	%
Communal area	16,355	41.9	16,355	41.9
Large-scale commercial	11,270	26.8	5770	14.8
Small-scale commercial	1400	3.6	2000	5.1
Resettlement	3090	7.9	7820	20.0
State farms/cooperatives	884	2.3	1000	2.6
National parks and wildlife	4900	12.5	4900	12.5
Forest areas	977	2.5	977	2.5
Urban	196	0.5	250	0.6
Total	39,072	100.0	39,072	100.0

Government's resettlement programme was therefore launched in a nexus of compromise, limited both by the pace at which commercial farm land could be acquired and by the ability of government to budget for and technically implement resettlement in the manner envisaged. Resettlement proceeded under four models:

(1) Model A involved village settlements with individual allocations of arable land of family and grazing entitlements of 4–10 livestock units depending upon agro-ecological zone;

(2) Model B, where commercial farms were to be converted into producer co-operatives;

(3) Model C involved the development of a nuclear commercial estate with outgrower producers; and

(4) Model D involved the provision of paddocked grazing areas, but not settlement, in neighbouring ranching areas.

In all cases occupancy was to be by permit, withdrawable if conditions were not followed.

Ambitious targets were set for the programme, the intention being to resettle 162,000 families (1,296,000 persons) by 1984. In the event resettlement under the programme proceeded at a far slower pace, particularly during the last 5 years of the decade. By the middle of 1989 a total of 2,713,725 ha had been acquired for resettlement (see Table 8.4) and a total of 52,000 families (416,000 persons) had been resettled, mostly under the Model A scheme. The reasons for this failure to meet earlier and more ambitious targets were economic, political, technical and institutional in character. The direct costs of resettlement along prescribed lines were far higher than originally estimated and the indirect costs in lost wages and

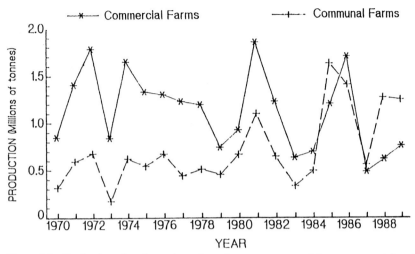

Figure 8.8 Production of maize and sorghum (tonnes per annum) from commercial and communal farming areas in Zimbabwe between 1970 and 1989

communal land earnings even higher. Kinsey calculated in 1984 that "the resettlement programme will, by the mid-1980s, be generating an annual loss in terms of wages and income foregone of some Z$ 83 million" (Kinsey 1984). These high costs, coupled with an economic recession over the period 1983–1985, made the government more vulnerable to conservative approaches to land reform espoused by the World Bank and Western governments and acted as a disincentive to vigorous implementation.

Continued high productivity in the commercial farming sector, now involving a significant number of Black elites, provided a further disincentive. In 1980, 42 per cent of the country had been held by 6000 White farmers. By 1989, the number of farmers had been reduced to 4319 (no longer exclusively White) owning 29 per cent of the land but still an effective lobby in political circles. Productivity in sections of the communal lands rose, with communal land farmers increasing their share of marketed food staples (see Figure 8.8) although their share of gross commercial sales has not increased proportionately (see Figure 8.9 and Figure 8.10). Even though the higher maize production on the communal lands involved comparatively few farmers in the higher potential areas, during the second half of the 1980s, all of this contributed to a diminished political immediacy over the issue of land reform.

Institutionally, the resettlement programme suffered from a number of disabilities, including a shortage of planning and implementing personnel and bureaucratic inefficiencies. More importantly, however, it was the technocratic and bureaucratic dimensions of the programme that impeded

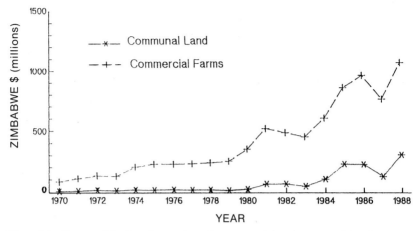

Figure 8.9 Commercial sales of crops from the commercial and communal farming sectors of Zimbabwe between 1970 and 1989

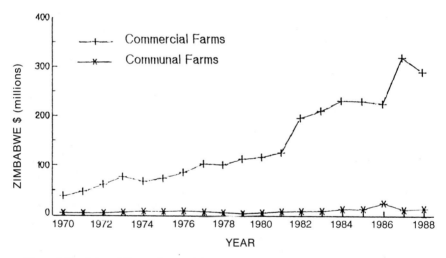

Figure 8.10 Sales of livestock and dairy products from the commercial and communal farming sectors of Zimbabwe between 1970 and 1989

its development, both in its planned implementation and in its acceptance by the peasant cultivators for whom it was planned. It was these dimensions which, for Drinkwater, were "the most important legacy of the colonial state", more important than "the racially drawn economic divide between Black and White". They:

"left the state at independence as the dominant source of power in the country and its generally centralised institutions in the habit of exercising this power through purposive–rational action. Those who work for state institutions are trained and socialized into the language of goals, policies, programs and plans, and hence accept as normal that bureaucracies should function according to a purposive rationality" (Drinkwater 1989).

This stance, combining centralised bureaucratic control with assumptions about appropriate technological strategies for agricultural development and resource utilisation, marginalises locally-evolved and locale-specific ecological knowledge, and inhibits the institutional dynamics necessary for effective communally-based regimes of resource management. Aside from the point that administrative approaches may be technically flawed because they tend to rely upon overgeneralised land classification and stocking rates, the resettlement schemes provided tenure arrangements that were neither attractive to prospective settlers, nor necessary for a long-term collective commitment to good husbandry. In these respects the resettlement programme differed very little from the technocratic policy applied to the communal lands during the pre-independence period before 1980.

Resettlement during the decade also involved large-scale migrations of people onto lands already classified at independence as communal lands but which had previously been sparsely settled. The Zambezi Valley and its escarpment areas were a particular target for this migration, receiving 60,000–80,000 immigrants during the period. In some cases this immigration was planned, the valley being perceived as an unexploited resource providing readily available land without any acquisition cost to government. The Mid-Zambezi Rural Development Project, financed by the African Development Bank and The Food and Agricultural Organization of the United Nations (FAO), provides an instructive example. In part the project resulted from the European Economic Community (EEC)-funded tsetse eradication programme, which was implemented without any land-use planning for the valley. Subsequently, FAO were brought in as government consultants to develop a land-use plan for the project area, which was elaborated broadly along the lines of the Model A resettlement scheme. It was planned to "rationalise" the settlement patterns and agricultural activities of the 4600 households estimated in 1986 to already be in the project area and also to settle 3000 new households there.

The project was therefore a resettlement scheme for an already settled area, combining the government's resettlement and agricultural reorganisation objectives. In this it was consistent with the First Five-Year National Development Plan for 1986–1990, which made these objectives clear:

In addition to the translocation resettlement which utilizes purchased former large-scale commercial farms, the re-organization of settlement

163

patterns in the Communal Areas will become part and parcel of the resettlement program. This entails replanning of land-use patterns in order to attain optimum exploitation of the agricultural resource potential on a sustainable basis and to ensure adequate provision of economic, social and institutional infrastructure (Republic of Zimbabwe 1986).

To date the project has had a difficult passage in implementation, suffering all the technical and institutional disabilities noted earlier for the resettlement programme. In a preliminary assessment Derman concludes that the project replicates the failures of top-down, highly centralised planning approaches that have been demonstrated elsewhere. Its most serious long-term consequence lies with its suppression of a local sense of proprietorship and a resultant sense of dependency. One of Derman's informants commented that "since the Government had taken responsibility for their lives it would be for Government to care for them when the project failed" (Derman 1990).

A further problem, which has emerged in the Mid-Zambezi project, is illustrative of the other form of resettlement which has taken place during the decade. Planned in 1986 to accommodate 3000 new migrant households, at implementation it was discovered that there were already more than the targeted 7600 households living in the project area; unplanned immigration had been ahead of government planning. In spite of official efforts to control it, voluntary and unplanned translocation into the valley and its escarpment areas has been the dominant mode of in-migration during the planning period. These migrants were a mix of land-poor peasant farmers from high-density communal lands, small-scale agricultural entrepreneurs and landless wage labourers retired from the country's industrial work force. A survey conducted in 1988 (CASS 1988) found that 70 per cent of these migrants had moved from communal farming activities elsewhere, but that among these only 59 per cent gave landlessness as their reason for moving. Among the remaining 41 per cent a major motivation was the acquisition of more land for cash cropping, particularly of cotton. There was also a significant proportion of migrants (26 per cent) who had moved from wage-labour sites in industry and commercial agriculture. Many of these were not only landless but also without any traditional claims or residential entitlement in any communal land. Most were either long-term labour migrants from Mozambique, Malawi or Zambia or the children of such migrants. People in the ex-wage-labour category were seeking a residential and subsistence base for their post-employment retirement years and figured prominently among those settler households clearing and cultivating slopes on the Zambezi escarpment. They are, to return to a point made earlier, a prominent example of the way in which communal lands are used to subsidise wage-labour structures in the economies that do not meet the post-employment and social security costs of "retired" people. In this particular case, it is the ecology of the Zambezi Valley which bears this cost, providing a further reminder that ecological sustainability is linked not only to rural land use and agricultural policy but also to industrial employment policy.

MACRO-ECONOMIC AND TRADE DEVELOPMENTS

At independence in 1980 Zimbabwe inherited a diversified but centrally controlled economy that was highly dependent on commercial agriculture for food security and foreign exchange earnings. The war of independence also brought in its wake the need for rehabilitation of the infrastructure, particularly in the communal lands. Government development priorities were properly directed to the rural areas, not only in agriculture but also in health and education. Some of the key macro-economic programmes influencing rural land use during the last decade included the Lomé convention and a variety of measures designed to raise crop prices.

Beef exports and the Lomé convention

The Lomé convention was established in 1975 between the EEC and 46 developing countries from Africa, the Caribbean and the Pacific, to provide selective duty free access to markets. The preferential prices offered for prime beef under the Lomé convention stimulated massive investment in the cattle ranching and beef industry. This involved the development of new abattoirs, stringent foot and mouth disease control measures including the fencing of corridors and large vaccination programmes, and an extension of the tsetse control programme into the Zambezi Valley. Resettlement programmes for the Zambezi Valley were facilitated by these developments.

Pricing and marketing measures for crops

Exchange rate controls and subsidies have stimulated the development of increased production of maize and cash crops in the communal sector. Increases in production are summarised in Figure 8.8 and Figure 8.9. Pricing policies have, however, continued to tax the producer and effectively maintain the imbalance between rural areas and the cities, by subsidising industry and city dwellers at the expense of the farmer.

Nominal government intervention in agricultural prices began with the

Maize Control Act and the establishment of the Grain Marketing Board in 1931, and has since steadily increased. Cotton marketing was brought under statutory control in 1936 and the Cold Storage Commission was established in 1937. In 1967 a new Agricultural Marketing Authority assumed responsibility for all marketing boards (Jansen 1982).

Following the Unilateral Declaration of Independence in 1965 and the imposition of sanctions, food security and foreign currency became primary political concerns. Governments have become increasingly concerned to maintain low food prices, and since 1976 many controlled food products have been subsidised either by the producer or from government revenue (Jansen 1982; Jansen, pers. comm., 1990). Guaranteed pre-planting producer prices were introduced in 1975 but abandoned after independence. Producer prices, however, continued to be revised upwards for some crops such as maize, as an incentive to producers and as a means of securing export revenue for other crops whilst government continued to subsidise consumer prices. Jansen (1982) concluded that pricing policy in Zimbabwe had achieved near self-sufficiency through incentives to commercial farmers. Consumers had also benefited but the subsidies had placed an enormous burden on Government revenue and, indirectly, on taxpayers.

Up to 1984 the only crop on which the producer had been effectively taxed was groundnuts. Since then there has been an implicit taxation of soyabean, cotton and beef producers, even if the overvaluation of the Zimbabwe dollar is ignored (see Table 8.5). At realistic exchange rates, however, all crops as well as beef cattle show a substantial tax on the producer during the 1980s (see Table 8.5). Subsidies to crops, dairy products and beef have increased from Z$ 46,000,000 in 1979 to Z$ 177,000,000 in 1990. The subsidies over the last decade have, in each year, amounted to between 46 per cent and 57 per cent of the Ministry of Agriculture budget. Since a high proportion of the subsidies are benefits to urban consumers and producer prices reflect a tax, in real terms, on producers, it is not surprising that small-scale commercial and peasant farmers (other than maize growers in some regions) have not prospered. Pricing policies have in practice increased the pressure on the land and almost certainly contributed to the popular perception that the real rural problem is "land hunger".

The problems facing rural peasant farmers are compounded by poor rural transport and marketing systems and, together with pricing structures, have led to adverse terms of trade in comparison with large-scale producers who have been able to diversify into other, higher-valued crops. In practice there has been little major change in the overall patterns of land use, production and environmental effects compared with earlier decades. The developments of the 1980s can probably best be summarised as "more of the same". The eradication of tsetse fly from large parts of the Zambezi

Table 8.5 Nominal rate of protection (NRP = ((Domestic price/equivalent world price) −1) × 100) for major crops grown in Zimbabwe at official and realistic exchange rates, 1981–1990. The beef NRP uses an actual sales weighted average of the export realisation and local sales realisation. Source: D.J. Jansen (pers. comm., 1990)

	Maize	Wheat	Ground-nuts	Soya beans	Cotton	Tobacco	Beef
At the offical exchange rate							
1981	17.4	−9.3	−18.4	22.0	−11.0	0.0	20.7
1982	84.8	−19.2	−11.9	0.0	31.0	0.0	27.5
1983	66.3	−10.9	−34.5	0.0	−2.0	0.0	7.4
1984	−45.1	−6.7	−34.2	0.0	−33.0	0.0	1.4
1985	35.8	9.9	23.5	0.0	−17.0	0.0	−8.4
1986	122.6	−0.5	−28.0	0.0	23.0	0.0	−14.3
1987	52.7	41.9	4.1	−45.0	7.0	0.0	−10.9
1988	−21.7	15.7	−0.1	−40.0	−13.0	0.0	−20.1
1989	−10.6	−20.1	14.0	−38.0	−5.0	0.0	−15.2
1990	12.9	−17.3	17.7	N/A	N/A	N/A	−19.7
Averages:							
1981–84	30.8	−1.9	−24.8	5.5	−3.7	0.0	14.2
1985–89	35.8	9.4	−4.9	−24.6	−1.0	0.0	−13.8
1981–89	33.6	4.4	−13.7	−11.2	−2.2	0.0	−1.3
At a realistic exchange rate							
1981	−9.9	−30.4	−37.3	−6.4	−31.7	−23.2	17.9
1982	23.0	−20.7	−41.4	−33.5	−12.8	−33.5	23.8
1983	9.0	−41.5	−57.1	−34.4	−35.7	−34.4	0.8
1984	−61.7	−35.0	−54.2	−30.3	−53.3	−30.3	−6.3
1985	−10.0	−27.2	−18.2	−33.7	−45.0	−33.7	−17.7
1986	37.6	−38.5	−55.5	−38.2	−24.0	−38.0	−25.3
1987	−9.7	−16.1	−38.5	−67.5	−36.8	−40.9	−27.1
1988	−51.9	−29.0	−38.7	−63.2	−46.6	−38.6	−33.6
1989	−49.8	−55.1	−57.3	−65.1	−46.6	−43.8	−19.9
1990	−41.8	−57.4	−39.3	N/A	N/A	N/A	−25.7
Averages:							
1981–84	−9.9	−31.9	−47.5	−26.1	−33.4	−30.4	9.1
1985–89	−16.8	−33.2	−41.6	−53.5	−39.8	−39.0	−24.7
1981–89	−13.7	32.6	−44.2	−41.4	−36.9	−35.2	−9.7

Valley has opened up more land to settlement, both planned and unplanned. The ecological characteristics and effects of subsistence agriculture have been extended without substantially relieving pressure on the densely populated communal lands. The basic agro-ecological question, which remains unanswered, is how can productive and sustainable agricultural enterprises be established so that they can cope with increasing human population densities.

CURRENT CONSTRAINTS AND CHALLENGES IN POLICY AND PRACTICE

By the middle of 1989 land ownership had resurfaced as a major political issue. The government was facing a national election in the following year, which would coincide with the expiry of the Lancaster House restrictions on the acquisition of private land by the state. The ZANU-PF Government thus made land one of its main election issues, stating that the current position was "morally unacceptable, economically unjustifiable and politically untenable" (Palmer 1990). Having won the national elections in April 1990 the government then went on to state its intentions to acquire an additional 5–6 million ha of commercial farm land for resettlement, most of it coming from Natural Regions II and III.

This intention was elaborated in a 15 point memorandum on land policy presented to Parliament by the Minister of Lands, Agriculture and Rural Resettlement at the end of July 1990. Among these points were the following:

(1) *Land acquisition.* The government would amend the Land Acquisition Act enabling it to acquire land when and where it was required for resettlement purposes, compensation to be "fair" but wholly to be determined by government and without recourse to the courts.

(2) *Control of land prices.* The government would control the price of agricultural land on the basis of variation in agro-ecological zones.

(3) *Land tax.* The government would impose a land tax based on the calculated production potential per ha per natural region. This tax would only be levied on the commercial farming sector.

(4) *Land inspections.* Land would be systematically inspected and land found underutilised over a period of time would be acquired by the state.

(5) *State farming.* The state farming sector would continue to expand and "consolidate its role as an agricultural producer alongside other large-scale commercial farmers".

(6) *Limitations on ownership and size of holdings.* Government would legislate against the ownership of more than one farm unit by one individual or company, stipulate minimum and maximum farm sizes by agro-ecological region and prohibit the ownership of agricultural land by absentee landlords or foreigners.

(7) *Settler selection.* In future, the government would select farmers for resettlement on the basis of proven agricultural competence. "In the future, all farmers should become commercial farmers".

(8) *Land tenure system.* The government would set up a commission to review current tenure systems and "recommend the most realistic tenure system for the country".

(9) *Structure of the agricultural sector.* The agricultural sector would continue to be comprised of the large-scale and small-scale commercial sectors as well as the communal, resettlement and state farming sectors.

In the ensuing debates the Commercial Farmers' Union (CFU) has argued that, while it accepts the need for land reform, this should be "implemented in a manner that ensures land is used on a sustainably productive basis". The Union estimates that the resettlement of 5.5 million ha of commercial farm holdings would have the effect of reducing agricultural production by Z$ 500 million and the value of agricultural exports by Z$ 425 million per annum. It also argues that the proposed methods of land acquisition will inhibit investment in private-sector commercial agriculture, and in a set of counter-proposals suggests that the "rate of resettlement should be matched to the resources available, to meet the aspirations of the people, but without jeopardizing national productivity". The CFU believes that "the lack of accountability for resources is probably the single most important contributory factor to their abuse" and recommends that in both resettlement and communal areas the government "should offer title to those farmers who have proved competent in utilizing the land on a sustainably productive basis" (CFU 1991). In response the Minister of Lands, Agriculture and Rural Resettlement has stated that the only debate that government is willing to entertain about the resettlement programme is on how it should be implemented and not on the need for such a programme. The government would try to produce a pragmatic policy that attempted to reconcile all the divergent political and economic interests encompassed by the land issue, noting that "we are a responsible government and we cannot afford to destroy the economic fabric that we have created" (*The Herald*, 12.1.91).

Assuming that government's intentions are implemented and that 5.5 million ha are transferred from large-scale commercial farms to resettlement land but with small additions to state farms, small-scale commercial farms and urban areas, one could posit a land-use classification scenario in the late 1990s as shown in Table 8.3 and Figure 8.1. Land under private commercial agricultural production would thus have shrunk from 30.4 to 19.9 per cent, with a further 17.6 per cent of the land being under direct state control, either as wildlife and forest estate or as state farms. Private commercial farm units would be diminished in size, and land use and production intensified. Where such reductions in farm size are accompanied by the full range of support systems (infrastructure, markets, inputs of equipment and energy, and fair prices for produce to the farmer) then adverse environmental effects are likely to be minimal. If the cost of intensification of land use, however, has to be borne by the land, then an expansion of the degradation (accelerated deforestation, erosion and declining productivity of arable land) that has characterised much small-scale and peasant agriculture in Zimbabwe can be expected. A key issue in the intensification of land use is the provision of adequate energy and infrastructure. Most resettlement schemes on former commercial farm land made no provision for the additional energy needs of settlers, with the result that deforestation was inevitable.

In addition, a number of economic, technological and implementational constraints are likely to inhibit government's land reform intentions at the levels projected. It can, however, be assumed that the 1990s will be marked by large-scale transfers of commercial agricultural land to other categories in the range of 2.5 to 3.5 million ha.

It is, however, with the greatly expanded area of land held under the communal land or resettlement categories that policy must be primarily concerned in terms of the criteria of economic efficiency, ecological sustainability and institutional viability. These are the challenges which policy must face if it is not to simply replicate the politically-driven short-term responses to land hunger which have characterised land-use policy and practice in Zimbabwe for over half a century. The reallocation of large areas of commercial farm land to small-scale producers might meet the immediate aspirations of the government's major constituency, but it should also afford the opportunity to improve the economic efficiency and sustainability of resource usage throughout the country and to fundamentally revise the economic and institutional structures necessary for this goal. If this is not done, the land reform programme of the 1990s will turn out to be no more than a replication, on a grander scale, of the incomplete and ineffective policies of land-use reform of earlier decades. Most Latin American attempts to pursue a similar policy path appear to have been abject failures (see Klink et al., Silva and Moreno, and Holmes, this book).

In their current form neither the government proposals, nor the CFU counter-proposals, fully address these problems. Both give some recognition to conservation and sustainability concerns and both emphasise economic efficiency in land use through more intensive management, agricultural competence and the provision of the necessary infrastructural services. But neither directly recognise that land and resources under communal regimes indirectly subsidise the national economy, nor do they recognise the opportunities provided to critically re-evaluate approaches that only assign conventional agricultural values to land. These opportunities include the redirection of resource exploitation strategies to take more account of international markets where Zimbabwe is well-placed to compete favourably. From this viewpoint the current emphasis on exports of cereals, cotton and beef seems inappropriate.

With more than 65 per cent of Zimbabwe's land area falling into Natural Regions IV and V and with limited potential for the development of irrigation, there is little likelihood of successfully using this land for commercial, export-orientated agriculture. Moreover, continuing use under high-density subsistence agriculture is unlikely to be sustainable. This area, like much of east and southern Africa, does hold a comparative advantage in the exploitation of indigenous large mammals. Some 2,700,000 ha of commercial farm land and 1,200,000 ha of communal farm land are presently being managed as multi-species wildlife systems for tourism, sport hunting and meat production, often in combination with domestic livestock. Preliminary indications are that this form of land use may be more economically productive than commercial cattle ranching, with lower environmental impacts (Child 1988). The intensification of wildlife-based tourism promises to provide far greater returns to the land, and in the western parts of Zimbabwe there is a marked swing to wildlife-based economies, both in commercial and communal lands. The potential in the region for this form of land use is enormous with about 17 per cent of the land area of the Southern Africa Development Coordination Conference (SADCC) region designated for wildlife utilisation (Cumming 1991b).

Perhaps the most critical gap in the current government land reform proposals lies in their failure to address the institutional components (land tenure and resource rights regimes, management options, motivational dynamics) necessary for efficient and sustainable land use in this category. Apart from the rather vague intention to appoint a commission to "recommend the most realistic tenure system for the country", nothing is said on this subject and one is left with the impression that, by default, the current situation will continue in which traditional and communal regimes are largely impotent and state technocratic controls are ineffective.

The CFU counter-proposals do address the issue in part, recommending individualisation of tenure for the farmer in communal and resettlement lands "by way of a lease convertible to title to enable him to develop his farm in confidence" (CFU 1991). This proposal, however, takes no account of scale, does not say how in a mixed production system involving cropping and livestock such privatisation of title would operate or how the grazing commonage would be managed. Furthermore, given the government's current ideological preference for an anti-individualistic socialism, it is unlikely that it will espouse recommendations for a thorough-going privatisation of tenure in the communal and resettlement land categories.

Between the CFU and government we have, therefore, two contrasting predilections for the management of resource exploitation on these lands: privatisation or state technocratic control. There is a "Hardinesque" quality to this situation, the available options conforming to Hardin's thesis that resources should either be privatised or controlled by central government authority to ensure sustainable use (Hardin 1968). Both perspectives ignore the disabilities encountered by privatisation or state control in the contexts concerned, and both ignore the demonstrated potential of communally-based common-property regimes to coalesce indigenous ecological knowledge with technological and economic change, motivate good husbandry, enforce collective conformity and collaborate with national concerns for ecologically sound and economically productive land use. All of these are institutional characteristics required for the implementation of a national policy at the variety of microenvironmental and context-specific levels exhibited in communal and resettlement lands, and without them the objectives of economic efficiency and ecological sustainability will not be achieved.

The policy basis for the active pursuit of this type of resource management regime in the communal lands already exists in the form of government's directives and legislation on decentralisation and local government. These envisage a "decentralization of the planning and supervisory functions of Government" and seek "a comprehensive and more effective system of involving the local communities both horizontally and vertically in the process of planning and effecting their development" (Murombedzi 1987). Pursuing these objectives the District Councils Act of 1980 established an administrative and planning structure for communal lands, incorporating a hierarchy of units running from village development committees to ward development committees to the district council. This structure has, to date, met with mixed success and has not in general achieved the desired result in terms of the devolution of initiative and self-sufficiency to local levels. There are a number of reasons, including the lack of administrative skills at local levels and a resistance in bureaucratic circles to any genuine devolution of authority. More importantly, however, the district councils, let alone the ward or village development committees, have not had the financial base for any genuine autonomy in planning and implementation. Without any significant tax base, and without any legal proprietorship over collective natural resources, they have been dependent on government grants for recurrent expenses and on government and aid funds for development projects.

Significantly, it has been the acquisition of proprietorship over the natural wildlife resource that has provided the catalyst for experiments by some district councils to break out of this syndrome of dependency. Using a provision of the Parks and Wild Life Act, which allows for the conferment of custodial use-rights over wildlife to district councils, the Department of National Parks and Wild Life Management has developed the CAMPFIRE programme which now extends to 12 of the country's district councils. Among the informing insights of the programme are the following:

(1) That wildlife should be promoted as an economic form of sustainable resource use for enhancing rural productivity in areas to which it is suited;

(2) That local communities and landowners are more effectively motivated to conserve wildlife when it is of direct economic benefit to them;

(3) That sustainable exploitation of the resource requires cause-and-effect relationships linking good husbandry with benefit;

(4) That proprietorship must include the authority to decide whether to use wildlife at all, to determine the mode and extent of its use and the right to benefit fully from its exploitation; and

(5) That in communal contexts the unit of proprietorship, with rights of inclusion and exclusion, should be as small as management considerations permit, allowing conformity to management regimes to be enforced by collective and informal pressure. The unit of proprietorship should be the unit of management and the unit of benefit.

The CAMPFIRE programme, having been implemented only in January 1989, is too young to be confidently cited as a success story. It faces a number of implementational problems, including bureaucratic resistance in certain circles and an inadequate legislative base. It has nevertheless already had a dramatic impact on the financial base of operations for certain district councils, council incomes having been sharply increased. In certain wards, producers now regard wildlife as an asset rather than a liability and "free rider" exploitation in the form of poaching has diminished. Communities have become more assertive over their claims to the proprietorship of all natural resources and have begun to make their own land-use plans to exploit and conserve the range of resources available. Collectively more confident about their own prerogatives and abilities, they are more conservative about the allocation of their land and more aware of the linkages that bind them to regional and national structures of resource exploitation.

All of this is indicative of the potential of communally-based resource management regimes to meet the institutional requirements of Zimbabwe's communal and resettlement areas. Such regimes make administrative and economic sense and provide for ecosystem and social system variation. They constitute an alternative to the "privatise or nationalise" debate over the allocation of land-use rights in Zimbabwe. It is not suggested that a common-property regime without any elements of state control or privatised tenure is appropriate for communal and resettlement lands. Different tenure arrangements for specific resources can exist within the same context, leading to a mix of individualised and co-management arrangements. What we are suggesting is that the tenure alternatives are broader than policy debates have indicated and that a common-property system with respect to certain resources has the potential to be institutionally viable, economically efficient and ecologically sustainable.

CONCLUSIONS AND RECOMMENDATIONS

From the analysis of Zimbabwean materials presented in this chapter, four generalised conclusions and policy suggestions can be made. These are stated below with specific reference to Zimbabwe, but are likely to be of relevance for savanna land-use policy in other countries with similar characteristics. Certain assumptions are made, namely that:

(1) Significant shifts in land categories will occur, transferring land under large-scale commercial production to small-scale agricultural production;

(2) No radical global economic recession will occur; and

(3) Global environmental changes will not radically alter the Zimbabwean environment.

Equity, economics and ecology

The gross inequities in Zimbabwe's dual agricultural and land-use system have far exceeded the limits of national social acceptability and acquired a political salience which makes them a dominant consideration in government policy. Economic and agricultural production considerations have, however, constrained policies driven solely by equity issues. The history of Zimbabwe has been marked by cycles of policy direction in which one or the other of these perspectives has been dominant, masking the conflicts involved and the trade-offs required. Ecological considerations have played an inconsistent role, at times being substantive but often being used as rationalisations for short-term political or economic imperatives. The current climate of heightened environmental awareness provides the opportunity for policy formulation that more candidly addresses the conflicts inherent in the equity–economics–ecology equation. This leads to our first policy prescription, which has to do with the policy process itself. Policy-making on land-use, on environmental issues and on the economy should be far less fragmented and disjunctive than it is at present. Policy-making should be more holistic and more synchronised. This will require a major coordinative effort on the part of government, with greater recognition of the political and economic compromises required and, particularly, more consultation at lower levels.

Macro-economic structures and policies

Essential for a viable and sustainable land-use policy is a macroeconomic environment that inhibits "subsidies from nature" for the interests of commerce and industry, for the benefit of urban populations and at the expense of rural land managers. Marketing and price controls contribute to this characteristic, as do wage policies that require rural areas to provide the social security requirements of the wage sector in the economy. International trade agreements and large-scale international aid projects, involving widespread technological and environmental impacts, may also contribute to an economic policy bias with this effect. Policy should therefore seek to redress this bias, giving special consideration to:

(1) The removal of adverse subsidies, price control and tax structures;

(2) The provision of wage policies in the formal economy which do not assume post-employment subsidies from the communal and resettlement lands; and

(3) A reassessment of international trade and aid agreements in the light of their long-term effects on the Zimbabwean economy and environment.

Diversification of natural resource utilisation

Two-thirds of Zimbabwe comprises semi-arid or arid savanna and the country can no longer afford to ignore the wealth-generating potential of diversifying its rural production systems to include the full range of sustainable indigenous natural resource utilisation. This range includes tourism, sport hunting, game farming and the presence of a number of economically valuable and exploitable species of flora and fauna not common elsewhere. Zimbabwe has a comparative advantage in these assets which has not as yet been fully exploited, largely because of conventional and commercially entrenched views on land use, reinforced by international aid and trade agreements.

Current constraints to the development of these diverse resources include outdated legal and bureaucratic structures, and in the case of some species the lobbying of anti-utilisation groups in the developed world. Zimbabwean policy should vigorously encourage the exploitation of these resources through innovative research and marketing, and revise legal and bureaucratic constraints where necessary.

Tenure and the institutions of land and resource management

Current policy debates on the tenure arrangements required for economic productivity and ecological sustainability are unduly restricted to privatised or state-control options. The potential for common-property regimes of management has largely been excluded from the debate but in certain contexts, prevalent in Zimbabwe, they offer the most promise for the sustainable and efficient use of the environment. They also provide a good institutional base for the diversification in resource exploitation advocated above. The primary constraints to developing this potential are entrenched technicist and bureaucratic perspectives, together with inhibitive legislative structures. Land management policy in Zimbabwe, particularly in the communal and resettlement lands, should therefore seek to build on, and give institutional and legal support to, generalised government intentions on decentralisation. This will involve not only a decentralisation of administration but also a genuine devolution of proprietorship and economic control and benefit as well.

ACKNOWLEDGEMENTS

We thank IUBS for inviting us to prepare this paper and for supporting our attendance at the Nairobi workshop. We are grateful to Meg Cumming, Jeremy Jackson, Ian Scoones and Michael Young for their helpful and critical comments on earlier drafts of this chapter. Mark Bowler kindly helped prepare the maps of land-use in Zimbabwe, and Doris Jansen generously provided information on crop pricing and nominal rates of protection.

REFERENCES

Arrighi, G. (1973) The political economy of Rhodesia. In: G. Arrighi and J. Saul (eds). *Essays on the political economy of Africa.* Monthly Review Press, New York.

Beach, D. (1977) The Shona economy: Branches of Production. In: R. Palmer and N. Parsons, (eds) *The roots of rural poverty in central and southern Africa.* Heinemann, London.

CASS (1988) A survey of in-migration to portions of the Guruve, Kariba and Gokwe Districts 1981–1987. Centre for Applied Social Sciences (CASS) University of Zimbabwe. Harare. (Unpublished ms.)

CFU (1991) Commercial farmers union. Proposals for land reform for Zimbabwe. Commercial Farmers' Union, Harare.

Cheater, A. (1990) The ideology of "communal" land tenure in Zimbabwe: Mythogenes is enacted? *Africa* 60:188–206.

Child, B.A. (1988) The role of wildlife utilisation in the sustainable economic development of semi-arid rangelands in Zimbabwe. Unpublished D. Phil. thesis. University of Oxford, Oxford.

Cumming, D.H.M. (1980) The management of elephants and other large mammals in Zimbabwe. In: P.A. Jewell, S. Holt and D. Hart (eds) *Problems in management of locally abundant wild mammals.* Academic Press, New York, pp. 91–118.

Cumming, D.H.M. (1989) Commercial and safari hunting in Zimbabwe. In: R.J. Hudson, K.R. Drew and L. M. Baskin (eds) *Wildlife production systems: Economic utilisation of wild ungulates.* Cambridge University Press, Cambridge, pp. 147–69.

Cumming, D.H.M. (1991a) Wildlife and the market place: A view from southern Africa. In: L.A. Renecker and R.J. Hudson (eds) In: Wildlife Production: *Conservation and Sustainable Development.* AFES misc. pub. 91–6. University of Alaska, Fairbanks. Fairbanks, Alaska, pp. 11–25.

Cumming, D.H.M. (1991b) Developments in game ranching and wildlife utilisation in east and southern Africa. In L.A. Renecker and R.J. Hudson (eds) In: Wildlife Production: *Conservation and Sustainable Development.* AFES misc. pub. 91–6. University of Alaska, Fairbanks. Fairbanks, Alaska, pp. 96–108.

Derman, W. (1990) The unsettling of the Zambezi Valley: An examination of the Mid-Zambezi Rural Development Project. Centre for Applied Social Sciences (CASS) Working Paper. University of Zimbabwe, Harare.

Drinkwater, M. (1989) Technical development and peasant impoverishment: Land-use policy in Zimbabwe's Midlands Province. *Journal of Southern African Studies* 15:287–305.

Grant, P. (1981) The fertilisation of sandy soils in peasant agriculture. *Zimbabwe Agriculture Journal* 78:169–75.

Grant, P. (1987) Soil problems associated with cropping in Natural Regions II, IV and V. In: *Cropping in the semiarid areas of Zimbabwe, Proceedings of a workshop held in Harare 24–28 August 1987.* Organised by Agritex and R & SS, Harare, pp. 588–91.

Hardin, G. (1968) The tragedy of the commons. *Science* 162:1243–8.

Harris, P.S. (1974) *Black industrial workers in Rhodesia.* Mambo Press, Gwelo.

Hughes, A.J.B. (1974) *Development in Rhodesian tribal areas: An overview.* Tribal Areas of Rhodesia Research Foundation, Salisbury.

Hussein, J. (1987) Agroclimatological analysis of growing season in Natural Regions III, IV and V of Zimbabwe. In: *Cropping in the semiarid areas of Zimbabwe, Proceedings of a workshop held in Harare 24–28 August 1987.* Organised by Agritex and R & SS, Harare, pp. 25–189.

Jansen, D. J. (1982) Agricultural prices and subsidies in Zimbabwe: Benefits, costs and trade-offs. Unpublished consultant's Report.

Kay, G. (1970) *Rhodesia: A human geography.* University of London Press, London.

Kinsey, B. (1984) The strategy and tactics of agrarian reform: Resettlement and land-use policy in Zimbabwe. Discussion Paper No. 136. School of Development Studies, University of East Anglia, Norwich.

Martin, R. B. (1986) *Communal Areas Management Program for Indigenous Resources (CAMPFIRE).* Department of National Parks and Wildlife Management, Harare.

Murombedzi, J. (1987) An outline of the historical trajectory of local government institutions and community development. In: *Zimbabwe Since 1923.* Proceedings of the Seminar on Issues and Experiences in Rural Development: China, Kenya and Zimbabwe. Centre for Applied Social Sciences (CASS), University of Zimbabwe, Harare, pp. 1–16.

Palmer, R. (1977) The agricultural history of Rhodesia. In: R. Palmer and N. Parsons (eds) *The roots of rural poverty in central and southern Africa.* Heinemann, London, pp. 221–254.

Palmer, R. (1990) Land reform in Zimbabwe 1980–1990. *African Affairs* 89:163–81.

Phimister, I. (1986) Discourse and the discipline of historical context: Conservationism and ideas about development in southern Rhodesia. *Journal of Southern African Studies* 12: 264–75.

Rappaport, R.A. (1969) Regulation of environmental relations among a New Guinea people. In: A.P. Vayda (ed) *Environmental and cultural behaviour.* Natural History Press, New York, pp. 181–201.

Schoffeleers, J.M. (1978) *Guardians of the land.* Mambo Press, Gwelo.

Scoones, I. (1989) Patch use by cattle in a dryland environment: Farmer knowledge and ecological theory. In: B. Cousins (ed.) *People, land and livestock.* Centre for Applied Social Sciences (CASS), University of Zimbabwe, Harare, pp. 277–309.

177

Stocking, M.A. and Elwell, H.A. (1973) Soil erosion hazard in Rhodesia. *Rhodesia Agricultural Journal* 70:93–101.

Vincent, V. and Thomas, R.G. (1961) *An agricultural survey of southern Rhodesia. Part 1 – Agro-ecological survey.* The Government Printer, Salisbury.

Whitlow, R. (1980) *Deforestation in Zimbabwe: Problems and prospects.* Supplement to Zambezia, University of Zimbabwe, Harare.

Whitlow, R. (1988) *Land degradation in Zimbabwe: A geographical study.* Natural Resources Board, Harare.

Yudelman, M. (1964) *Africans on the land.* Harvard University Press, Cambridge.

Zimbabwe, Republic of (1982) *Transitional National Development Plan.* Government Printer, Harare.

Zimbabwe, Republic of (1986) *First Five-Year Development Plan: 1986–1990.* Government Printer, Harare.

CHAPTER 9

RANGELAND BIOECONOMICS IN REVOLUTIONARY SOUTH AFRICA

M.T. Mentis and N. Seijas

SUMMARY

This study reviews the influence of government policy on economics, ecology and social equality within the savannas of South Africa. Policy and legislation have created two systems, built on a premise of separate development of Blacks and Whites (apartheid) where commercial and subsistence pastoralism co-exist. The intention of this apartheid system was to confine Blacks to relatively small areas of land, to restrict their movement to prevent market access and thereby protect Whites from competition. Policy and legal instruments to achieve this date to the colonial era, although the measures were strengthened with the rise of Afrikaner nationalism. Circumstances are now favourable for reform. Demonstrably apartheid, and central government control associated with it, are economically, ecologically and socially unsound, to the degree that the lives of even those that apartheid sought to benefit have been traumatised.

The case is a microcosm of a bizarre experiment in social engineering. The experiment holds several lessons relevant to renewable resource management. Strong central control is an object lesson in unintended consequences. Intervention has awarded privileges differentially, distorted market forces and subsidised overgrazing. Research and advisory services can be privatised, probably with great cost-effectiveness. The question that remains to be answered is: is it sufficient to repeal the laws that prevent Blacks from buying White land, or is it necessary to accompany these reforms with new forms of land taxation and even the redistribution of land?

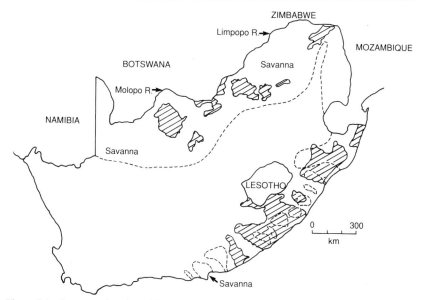

Figure 9.1 Savannas in a broad horseshoe shape around the north, north-east and south-east of South Africa. Subsistence farming areas are cross-hatched; the remainder are commercial farming areas

INTRODUCTION

This case study reviews the influence of strong, centrally controlled government policy on the savannas of South Africa. That policy includes apartheid legislation, central economic planning and national conservation policy. The study area includes 386,000 sq km of savannas in a broad horseshoe shape, south of the Molopo and Limpopo Rivers and down the east coast of South Africa (see Figure 9.1).

History and political context

The year 1652 was a turning point in the ecological history of South Africa: it was the year Whites arrived. Prior to that, the region was inhabited by Black tribal pastoralists. When Dutch merchants settled on the southern tip of the continent, in what is now Table Bay in the city of Cape Town, an era of conflict over control of the land began. Since then, successive colonial and modern South African governments have intervened in the land transactions between Blacks and Whites, taking stronger and stronger control over property rights and land use.

From 1652 to 1815, the Cape Colony remained under Dutch colonial rule. But in the wake of the French Revolution, as French armies swept over the Netherlands, the Dutch could no longer protect their colonies. They consented to British occupation of the Cape, to keep it from falling into French hands. In 1815 the British paid the Dutch £6 million and kept the Cape.

Dutch, or Afrikaner, resentment of British imperial rule was fierce. By the end of the century, they went to war. In the Anglo-Boer War of 1899–1902, the Afrikaner boers (farmers) lost. By this time South Africa had grown to four individual colonies, which were united under British rule in 1910.

One of the first South Africa statutes after Union was the 1913 Native Land Act, under which Blacks were permitted to own land in only 8 per cent of South Africa. (This was extended to 13 per cent under the Native Trust and Land Act of 1936.) The 1913 Land Act also prohibited Black rental farming (Black commercial farming on White-owned land in return for an annual rent) and sharecropping between Blacks and Whites. The major purpose of the Act was to cut off Black farmers' source of independent wealth, and force them into wage labour for Whites, especially in the mining industry.

Black leaders protested – the South African Native National Congress, later called the African National Congress or ANC, was formed in 1912 – but even a deputation to the British government failed. For the next 35 years, more racial laws followed. In 1948, the Nationalist Party came to power on a platform of grand apartheid, or "separate development", and South Africa then entered an era of centrally controlled social engineering.

Apartheid is of long standing. The first recorded racial segregation measure was the planting of a hedge of bitter almonds in 1660 to demarcate 6000 acres of the Cape Peninsula, dispossessed from the Khoikhoi. The chief architect of official apartheid policy was a Dutch immigrant named Hendrik F. Verwoerd who, as Prime Minister, created South Africa's nine tribal reserves in 1959 and later made South Africa independent of British rule. Over the next 20 years, more than three million Blacks were forcibly relocated to the tribal reserves, or "homelands" – remote, undeveloped regions included in the 13 per cent of South Africa's land relegated to Blacks, and ruled by "tribal" authorities, fostered and given inflated powers by Verwoerd.

Today, four of the original tribal areas are nominally independent, but their governments are heavily dependent on the South African government for financial support. The other five reserves are officially described as "self-governing", but the South African government retains various legislative controls over their policies.

This is the legislation that circumscribes conservation and land-use policy in South Africa today. As a result of the Land Acts and the homeland policy, two pastoral systems co-exist within South Africa. The first system is an essentially wealthy, White, commercial agricultural economy, where Whites own the individual farms and ranches, heavily regulated by government. The second system is characterised by a generally poor, Black, traditional or subsistence economy in tribal areas, where land is owned by tribal authorities (see Figure 9.1).

The government of President F.W. de Klerk, however, has fully accepted the untenability of apartheid, the homeland system and White minority domination. There are now prospects for liberalisation and democratisation. It is at this time of hope and uncertainty that we review the effects of South Africa's experiment in central planning and social engineering on rangeland conservation.

Types of land use and distribution of resources

The major land use of southern African savannas is extensive commercial (280,000 sq km) or subsistence (70,000 sq km) pastoralism (see Figure 9.1). The average White commercial ranch is about 2500 ha in size, while Black subsistence units are about 1 per cent of that size. Cattle and goats are the main domestic livestock types, managed for meat production. Private game reserves and ranches have been displacing commercial cattle ranching since the 1960s. Nationally and provincially administered game parks occupy some 34,000 sq km. Tropical and subtropical fruits, sugar-cane and annual crops are limited to regions where soil and rainfall permit, or where irrigation is possible.

ECOLOGICAL CONSTRAINTS TO DEVELOPMENT

Climatic and soil limitations

Although generally semi-arid and moisture-limited, South African savannas range from moist to arid. The moist regions in the centre and east, occupying 116,000 sq km with rainfall of 500–1000 mm per annum, have dystrophic soils. The main limitations on land capability here are not moisture but steep-slope and shallow soils. The arid regions of 270,000 sq km, mostly in the extreme north and west and with a rainfall from 250 to 650 mm per annum, have meso- or eutrophic soils. Land capability is limited by low and erratic rainfall.

Current condition of the study area

White-owned commercial ranches are widely regarded to be bush-encroached, as a consequence of heavy use of the herb layer and its displacement by a relatively unpalatable shrub layer. Black-owned subsistence land is less severely encroached because heavy stocking with goats coupled with a high human population density with a high demand for fuel wood has meant that bush encroachment is less of a problem.

Soil loss is severe in some of the semi-arid savannas. The national annual loss (as sediment in rivers) is estimated at 300–400 million tons, or about 3 tonnes per ha per year. There is local variation. Cultivated land, of which there is some in savanna, experiences severe soil loss. In the arid savannas rainfall erosivity is low and the erosion hazard is minimal, especially to the west where soils are sandy and relief low. In the moist savannas with generally favourable growing conditions for plants, a protective vegetal cover of unpalatable grasses is maintained despite heavy grazing. In the intervening semi-arid regions erosive convective thunderstorms prevail, and the climate has been conducive to the genesis of highly erodible soils. Coupled with high relief, spectacular soil loss has occurred.

Relation between conditions and productive potential

Moist savannas have an average herbaceous production of 2000–3000 kg/ha. Woody biomass is typically in the region of 20,000 kg/ha. These parameters, however, vary locally, depending on soil conditions and on land-use history. Sodic soils, for example, have low bush biomass and virtually no herbaceous production. Vertisols are more or less treeless. Herbage is typically "sour" (loses nutritive value upon maturity). Above 700–800 mm per year the herb layer responds to heavy grazing by a change in species composition from palatable to unpalatable grasses. As a result, there is frequently sufficient fuel in the dry season to carry intense fire, which can reduce woody biomass. At lower rainfall, heavy grazing causes denudation and accelerated soil loss, after which there is rarely sufficient fuel-accumulation to carry fire, with the consequence that bush encroachment can be a severe problem in low rainfall areas.

Average herbaceous production in arid savannas ranges from 500 to 1000 kg/ha. Woody biomass is about the same as for moist savannas. Herbage is typically "sweet" (grasses maintain nutritive value). Heavy grazing leads to denudation which, at the high rainfall extreme, is followed by accelerated soil loss. Maximal herbaceous production occurs at low or modest bush density, but decreases exponentially with increasing bush density. The herb layer under bush canopies is relatively "sweet", but has a low production. Between canopies the production is high but relatively "sour". Bush-clearing may increase herbaceous production but at the expense of lowered nutritive value.

SOCIAL CONSTRAINTS TO DEVELOPMENT AND CONSERVATION

South Africa is often described as both a "First World" and "Third World" country, terms euphemistically used to mean "White" and "Black", respectively. The traditional, tribal customs of rural Blacks are often pejoratively described as Third World traits, and are seen to be the major social constraint to development and conservation. We deal first with the Black system, then the White system and lastly the resource and conservation implications of removing the social and economic barriers that characterise them.

The Black social system

It is widely but incorrectly held that the social values of rural Black South Africans are irrational, and lead to waste and destruction particularly of the grazing resource. Traditional African pastoralists attach several different values to livestock. Primarily, they regard cattle as wealth in themselves, and attach variable importance to slaughter and milk value, sale or barter value and ceremonial and symbolic value. The number of animals an individual grazier possesses is usually seen as far more important than perceptions of breed quality or productivity.

Traditional Black cattle owners therefore maximise animal numbers. Development aid administrators and conservationists typically conclude from this that traditional graziers behave without regard to the ecological constraints of the land. They are perceived by some to have no understanding of long-term planning, preferring to increase herd sizes indiscriminately until pasturage is denuded (but see chapter 3 by Scoones et al.). Over the past half century there has been a persistent litany of warnings that, in tribal areas, the grazing resource is on the brink of collapse (Tapson 1990).

But the available evidence, both theoretical and empirical, does not support these claims. Given that the traditional pastoralists do maximise their livestock numbers, livestock density should settle at Caughley's (1976) "ecological carrying capacity", that is, the vegetation–animal equilibrium that the land can bear. This ecological carrying capacity is far greater than what many First World pasture experts consider the "economic carrying capacity", or the optimum number of livestock units for efficient commercial production (e.g. maximising return on investment). Experts who judge tribal land in terms of its capacity for commercial production would therefore deem it overstocked and in poor condition, long before it had reached its ecological capacity. In other words, they would underestimate the actual, physical limit of the land because they do not recognise the social constraints and imperatives.

The model of Jones and Sandland (1974) provides a useful framework in which to consider the above. According to the Jones–Sandland model, individual animal performance declines linearly with an increasing stocking rate. The curve for total animal production per unit area is an inverted parabola, intersecting with the horizontal stocking rate axis at zero and again at the point where stocking rate is equal to animal maintenance (i.e. where the animal performance per animal curve intersects with the stocking rate axis). "Economic carrying capacity" is about one-third to one-half of the stocking rate at maintenance, according to the economic formulae devised by Edwards (1981). This assumes access to credit at the same rate as White South Africans, the same market prices and the same ability to afford veterinary services. But Black South Africans do not have this. As a result, one would expect traditional pastoral systems to be stocked at two to three times the economic carrying capacity. Empirical data shows that this is often the case (Danckwerts 1981; Mentis 1981), taking into account that one White-owned beast might be bigger than one Black-owned beast (see Figure 9.2).

Not only do the Jones–Sandland and Caughley models explain these phenomena very well, they suggest two further predictions regarding ecological carrying capacities. Empirical evidence supports them both. First, according to the models, had tribal pastoralists been destructively overstocking their land, we should observe in South Africa undue stress applied to the veld, more or less irreversible range deterioration and, ultimately, a decline in stock numbers. In the tribal area of KwaZulu, however, stock inventory shows relative stability over the past 15 years, rising from 1.27 million in 1974 to 1.53 million in 1988. Minor stock increases during the period coincided with favourable seasons, and were accompanied by a decline in mortality (Tapson 1990).

Similarly, stock numbers in the Ciskei and Transkei tribal areas have been stable over an even longer period. From 1952 to 1980 stock numbers increased marginally from 1.45 million to 1.6 million (Bauer 1983). Periodic declines coincided with droughts (and not with rising cattle prices), and increases occurred following favourable seasons. Bauer comments: "The major implication of this is that there is no economic limit to the number of cattle placed on ... grazing land in the Transkei and Ciskei but there is an ecological limit, and it seems to have been reached" (Bauer 1983). According to Tapson, the traditional pastoral systems have "stubbornly

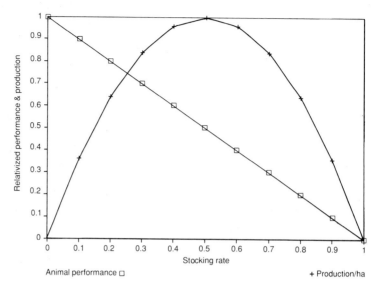

Figure 9.2 In terms of the Jones–Sandland model, (a) animal performance (□) is an inverse linear function of stocking rate, and (b) animal production per hectare (+) is an inverted parabolic function of stocking rate. Units have been relativised. Stocking rate is expressed as a fraction of the stocking rate at animal maintenance. Animal performance (e.g. mass gain per animal per day) is converted to a fraction of maximum animal performance. Similarly, animal production per hectare is expressed as a fraction of maximum production per hectare

resisted complying with the prediction of imminent collapse" (Tapson 1990).

The second prediction suggested by the models is that artificially supporting livestock production by creating additional waterholes, providing drought relief, disease control, importing high-production sires and so forth (Mentis 1986) would disturb the natural vegetation–animal equilibrium because this makes it technically feasible to increase the number of animals in a given area and put unprecedented pressure on the vegetation and promote range deterioration. In accordance with this theory, Bauer observes that various government schemes during drought years in Transkei and Ciskei have interfered with the natural dynamics in carrying capacities, "thus helping to prevent natural rejuvenation of pasturage" (Bauer 1983).

Although not well documented, many examples of range deterioration can be seen in South Africa where natural waterholes have been augmented, in wildlife reserves as well. Wildlife managers have, paradoxically, supplied additional watering holes, observed severe impact on the vegetation and then instituted culling of wild herbivores.

Further, a relatively new economic theory – the "farm-household" theory – provides a much more satisfying, if not predictive, explanation of the behaviour of rural families than does neoclassical economics (Low 1986). According to this new theory, the household is a production/consumption unit which tries to combine market inputs with the time of its members so as to satisfy consumption requirements as cheaply as possible. With the rise in urban and labour wages, households have changed their activities from purely working the land, tending the herds and performing domestic chores to include activities that produce money. They have increasingly split their efforts between high-income employment in the urban commercial sector and rural domestic activities. The labour that was formerly available for animal and crop production has been reduced, and farm production has declined. The conclusion is (1) that families prefer wage-earning, so long as they can *purchase* more consumption goods than they could *produce* themselves at home in the same amount of work time; and (2) that home activities are now focussed on the production of non-commercial goods and services, such as own child-care.

In summary, the following general points can be made regarding traditional pastoral systems. First, traditional pastoralists behave quite rationally given their particular values and circumstances. Second, the government intervention in the traditional pastoral systems has been wasteful at best, and, perhaps more realistically, harmful. It has focussed only on the production side of what is recognised by farm-household economics as the dual production/consumption nature of the rural household. That is, government activity has concentrated on improving home-farm production, ignoring the tendency of family members to substitute home-produce with a wage that can buy the goods they need. Third, the ecological integrity of the traditional systems has persisted despite intervention from government, bearing testimony to the extreme resilience of savanna systems to perturbation.

Land tenure

It is also commonly believed that South African tribal areas suffer from a "tragedy of the commons" (Hardin 1968): tribal land tenure is thought to be communal, where all individuals own all the land, giving none any incentive to conserve the land he or she uses. Although some tribal customs may give that impression, a closer analysis shows that is not the case. Writes Transkeian author Digby Koyana:

> [T]here is a private-law limitation – the *buqisa* custom. Under that custom everybody's livestock is allowed to graze on everybody's unfenced arable lands after reaping has already been done. That period, known as the *buqisa* season, takes place during the months of June, July and

August and even September when the old mealies have been reaped and no new ones have been planted ... Once crops are planted in spring, buqisa season is over and stock-owners are obliged to prevent their stock from trespassing on the cultivated lands, otherwise they become liable in damages for trespass (Koyana 1980).

Current land-tenure rights in KwaZulu further expose the "tragedy of the commons" as a fallacy in tribal areas. Tapson observes that there are strict grazing rights in practice in KwaZulu, which fall into two groups: specific pasture land designated for public use, and private land made available for public use. Small grassland plots that are part of a family landholding are not available to the public. "Only on the formally declared public pastures do the open-access rules apply which give them the 'communal grazing' character. Even these are converted to fields owned by individuals over time" (Tapson 1990). According to custom, those who have access to land retain their usufruct, provided they continue to use the land productively (Low 1986). Thus, contrary to current perceptions, there appears to be no widespread, unregulated "commons" in many tribal areas. There is a relatively sophisticated system of grazing rights that promotes efficient land use, enforced by custom and tradition rather than legislation.

Interaction with other markets

The history of interaction between Black and White agricultural markets provides further evidence of the rational behaviour of tribal farmers and their understanding of the ecological limits of land. Prior to the Land Acts of 1913 and 1936, sharecropping and rental farming between Blacks and Whites were widespread. They were an important source of independent wealth for Blacks, but also a source of heated opposition from a powerful lobby of White farmers, who were starved of servants and cheap farm labourers. Toward the end of the century, various colonial governments passed legislation to restrict Black land ownership and Whites' rights to employ Blacks as rental farmers and sharecroppers (which later became part of the provisions of the 1913 Land Act). Coupled with natural population increases, such restrictive legislation helped create a land shortage. Naturally, the incidence of relative denudation of the land by Blacks, faced with a shrinking grazing resource, increased. In the Transkei, for example, natural population increases were augmented by the influx of Black farmers pushed off the land elsewhere in the Cape Colony by the Location Acts of 1892, 1899 and 1909 (Bundy 1988).

As a result, many tribal pastoralists resorted to transforming fallow bushland in the process of natural regeneration into additional garden plots for grazing, and clearing trees and shrubs to increase grazing resources. "Black peasants were compelled to exist on a smaller total arable area, so that normal rotation was disrupted, fertility lessened, erosion hastened and returns diminished" (Bundy 1988). Where land was still available and the law permitted, however, Blacks bought and leased land to increase available grazing resources. In the wake of the discovery of gold in the Transvaal in 1886, for example, many White landowners went gold prospecting and more land became available for Blacks. One tribe leased 22 White-owned farms for grazing purposes. Another bought 18 farms and another 11 (Bundy 1988). The pressure of population on the land Blacks purchased was less than half of that on government-owned Black 'locations'.

Distribution of rights to use and develop resources

Prior to the Land Acts of 1913 and 1936, when Blacks and Whites competed for land relatively freely, those who did not use land efficiently – those who overstocked the pasturage and abused the rangeland – tended to sell or lease it to those who did conserve properly. They could not make a living on the land in such a competitive market if they did not use the grazing resource economically. Black tribal farmers frequently bought land from Whites. In Natal, for example, 70,000 acres were purchased by Black farmers between 1881 and 1890, and nearly 60,000 were bought in the next year alone. Indeed some White-owned land companies in the Transvaal in the 1880s had a policy of leasing land only to Blacks, who earned greater profits and paid better rents than Whites (Bundy 1988).

The Land Acts destroyed this market mechanism, by forbidding Blacks from buying and renting land from Whites throughout most of the country. Competition for land was restricted and more efficient Black farmers were prevented from displacing less efficient Whites, and the incentive for Whites to increase their efficiency was reduced. More land also became available to Whites than would have in a freely competitive market. One can assume that the costs to individual White landowners of abusing the land went down. White farmers could afford to abuse rangeland far more, without exhausting their available grazing resources and without losing significant agricultural profits.

Land-use regulation

There are several kinds of regulations governing the ploughing of virgin soil, the burning of veld and stocking rates. Several Acts of Parliament cover soil conservation and protection issues, but there are currently two major central government strategies designed to protect savanna lands. Both are fundamentally problematic. One is the National Grazing Strategy, a set of national requirements, including a set number of livestock units permissible per hectare in different regions of the country, farmers must meet in order to qualify for drought-relief aid. But it is well known that the appropriate number of livestock units for a particular farm depends largely on the grazier's objectives, his system of management, the starting state of the system, his financial position and the state of the economy (Mentis 1984). These influential factors are highly individualised and dispersed throughout the agricultural economy. Data on individual graziers are too diverse and too rapidly changing to be collected by one central body and reflected accurately in a national policy. By nature, then, the National Grazing Strategy recommendations are based on insufficient data, and thus are likely to be far less efficient for any given individual farmer than his own judgement, based on his knowledge of his own circumstances augmented by available agricultural research.

Nevertheless, to qualify for state drought-relief subsidies, farmers must follow the national strategy, even when it might conflict with their own assessment of proper herd size. This disincentive to use individual judgement increases the likelihood of inefficient pasture use. It also reduces the propensity of farmers to innovate and experiment with new grazing strategies.

The second major centrally-imposed land-use regulation is the Subdivision of Agricultural Land Act, passed in 1970. Under the Act, landowners are forbidden to subdivide certain agricultural land and sell it as smaller tracts. The intent of the Act was to prevent farmers from dividing their land into uneconomic units, unable to support enough livestock for the individual farmer to survive economically. But the unintended consequence of the Act is to further reduce competition for land. Outlawing the division of land into smaller, less expensive units cuts many lower- and middle-income buyers out of the market. Also, would-be small plot owners are compelled to buy big plots which, because of inadequate capital, they then use inefficiently. Unrestrained competitive market forces would naturally adjust property sizes to economic units, especially in those savanna regions where production potential is high and smaller units would be economically viable. Instead, the restriction of competition under the Act further reduces the incentive to use land efficiently.

The ill effects of subsidisation can be demonstrated further by analogy to the bioeconomic modelling of fisheries. According to Clark (1985) subsidies paid to fishermen reduce the cost of resource exploitation, prompting faster or greater resource depletion. By analogy, direct and indirect financial support to graziers by the state may be expected to lead to subsidised overgrazing.

Systems of enforcement

The National Grazing Strategy is generally enforced through the promise of drought-relief subsidies for those who follow it. Subsidies, however, even in times of disaster drought, would seem to discourage farmers from managing rangeland efficiently. They reward the inability to adapt to changes in circumstances of the market.

The subsidisation process has also been open to abuse. Until recently the requirements of the National Grazing Strategy have been enforced on a regional basis, with regional soil conservation committees carrying the responsibility for the evaluation of the relief needs of farmers after drought seasons. Subsidies have been distributed to entire regions, allowing individual farmers to benefit from drought relief in their local region regardless of whether their individual circumstances were caused by actual drought or their own misuse of the grazing resource.

As of June 1990, however, drought-relief is now distributed on an individual basis and only to those farmers who can prove, to the satisfaction of the Department of Agriculture, that they have complied with the national strategy. Moreover, farmers must now register with the national Director of Financial Assistance before a drought occurs, in order to qualify for drought relief in future. Records of compliance with national recommendations for at least 12 months prior to a drought must be submitted to the director before relief funds will be paid.

While this method may help contain flagrant abuses of drought-relief subsidies, it does not solve the fundamental problem that it rewards inefficient management, discourages farmers from maintaining production potential and rewards a lack of adaptability. The revised scheme still provides a disincentive for farmers to plan ahead for times of severe drought; it still motivates farmers to rely on national recommendations, which are by nature based on contested theories (Mentis et al. 1989) and on insufficient information about each farmer's individual circumstances. It also discourages individual innovation.

Agricultural policies and programmes

The sale of major agricultural products in South Africa is subject to numerous Acts as well as 22 statutory control marketing boards. Most pertinent to rangeland conservation is the Meat Board. The red meat market in South Africa is divided into nine controlled areas, which include all the major city centres within the country. In the controlled areas livestock farmers must direct all livestock to a specific, licensed abattoir.

Abattoirs serving controlled areas are, with one exception, owned and run by a statutory abattoir organisation which prohibits competition in price and services between individual abattoirs. In controlled areas, livestock producers are also prohibited from selling stock directly to retail butchers, but must sell all stock in a manner prescribed by the Meat Board at a specific, licensed auction market.

The additional functions of the Meat Board within the controlled areas include:

(1) Registration of all red meat producers and use of their compulsory annual returns of livestock numbers for purposes of planning;

(2) Regulation of the supply of slaughter livestock to controlled area abattoirs, through permits issued to individual producers and quotas for amount of stock to be slaughtered;

(3) Setting of the minimum guaranteed price for livestock;

(4) Setting of standards for quality of meat and of flaying and dressing practices;

(5) Grading of all carcases according to prescribed grading standards; and

(6) Trade in offal and hides.

The stated aim of such strict control over the red meat market is to help insulate livestock producers from the inherent risks of the agricultural market, and to help "stabilise" consumer prices. In terms of rangeland conservation this is a dangerous practice. For example, the minimum price set by the board prevents producers from reducing their livestock numbers quickly, at the onset of a drought, to cope with the diminishing grazing resource, as it does not allow prices to fall far enough to entice consumers to buy up the surplus of red meat on the market in the drought's early stages. While this may "protect" farmers from a drop in prices as weather conditions worsen, it often forces them to hold more livestock on their land for longer periods of time during a drought period than they would have, if prices were permitted to fall. This often has disastrous consequences for the producers themselves as well as their grazing resources. During the late stages of the drought period of 1982–83–84, for example, several herds of roughly 500 livestock units in the Western Cape region were reduced to 20 or 30, indicating several pasturages in the area were totally denuded. According to the independent Organisation of Livestock Producers, "the surplus removal floor price scheme by the Meat Board is the reason that in times of oversupply (drought), market access is restricted. This floor price scheme does not relieve congestion, it compounds it" (Organisation of Livestock Producers 1988).

Stringent meat hygiene regulations also have a destructive general effect as they make it difficult and expensive to market venison (although the control is not as strict as for beef and mutton). As a consequence, many cattle and sheep producers who also run antelope have little incentive to market antelope meat. The antelope numbers rise and subsequently compete with cattle and sheep for the grazing resource.

If hygiene regulations as well as price controls on red meat were lifted or relaxed, market prices would reflect what consumers are willing to pay for beef, mutton and venison at less stringent or varied standards of hygiene. These regulations were indeed lifted in Ciskei, and neither consumers nor producers have objected. A freer system could well make available an abundant supply of cheap meat (venison) to low-income consumers who currently can rarely afford red meat.

Trade-related policies

International sanctions on South Africa have created barriers to the export of livestock, which potentially reduces the ability of farmers to reduce livestock numbers quickly at the early stages of drought. With the approval of the Minister of Agriculture the Meat Board determines an annual quota for the total export of red meat. A portion of that quota is reserved for the board itself, to export some of the surplus meat it has bought from livestock producers at floor prices and only limited access is given to livestock producers. More important to the consumer is the control the Meat Board exercises over the import of cheaper meat from neighbouring Botswana, Namibia and Zimbabwe. As foreign competition is limited, production within South Africa is inefficient and, hence, consumers must pay higher prices and effectively subsidise production within the country.

Macro-economic policies

Allowable tax deductions for commercial farmers are generous. Individual farmers (unlike other taxpayers) may average out taxable income from farming to "equalise" tax payments in the face of annually fluctuating income (a result of the erratic and poorly predictable rainfall). Accelerated depreciation arrangements apply to machinery. Taxation arrangements for livestock replacement following drought appear to make it optimal to avoid destocking early. The effect is an incentive to put greater pressure on the grazing resource in dry and drought seasons than farmers would normally do in a freer market.

A state-run Land and Agricultural Bank was inaugurated in South Africa in 1912 to supplement state assistance to White farmers. The Land and Agricultural Bank, commercial banks, farming cooperatives and the Department of Agriculture have since become important sources of loan capital for Whites. Short-, medium- and long-term loans are issued against security of agricultural property or a promissory note. Real interest rates have been low or nil and, when necessary, have been subsidised by government. In addition to offering cheap loans, government has also come to the rescue of White farmers with drought aid and stock reduction schemes, while conservation programmes have extended the easy taxation arrangements described above.

This policy of "easy credit" has protected inefficient White farmers and completely excluded Black farmers. Blacks have not had comparable access to capital, assistance and aid under these generous terms. Denied property ownership rights, they have been unable to use land as security for a loan. As a result, farming production has almost certainly been lower and less efficient than it would have been in a more freely competitive market. By 1989, total farmer debt (comprised only in part by ranchers in the savanna) approached R15 billion, against a gross value of agricultural production around R18 billion.

To combat their rising debts, commercial ranchers have resorted to several tactics. First, they have gambled with erratic weather, cultivating marginally arable land and stocking at high levels in the hope of good rain and bumper production. While Danckwerts and King (1984) have demonstrated that conservative stocking yields a greater long-run profit, South African tax and credit policy incentives tend to encourage and reward exploitative stocking regimes, with the consequence that land degradation is greater than it otherwise would be.

Second, financially desperate ranchers have resorted to annual crop production in regions marginally suited to such practice. Third, when these desperate measures have also failed, annual cropping has been abandoned. The old fields have become vegetated by pioneerish plants of lesser productive and nutritive value than the native pastures.

From 1970 to 1987, the South Africa Reserve Bank increased the monetary base nearly eightfold and severely debased the South African rand. This has had several destructive effects. First, it has caused a sustained period of high inflation. Between 1970 and 1987, the cost of a basket of consumer goods rose nearly eightfold (Caldwell 1989). The current inflation rate is roughly 15 per cent per year. This continual decline in the purchasing power of money creates a strong disincentive to save and also to consider the long-term implications of investment decisions. As a result White commercial farmers are far less likely to prepare themselves financially for dry seasons and droughts. The high inflation rate also creates a greater demand for land because it retains its real value during inflationary times.

In turn, this increased demand for land drives up the land price and places it further out of the reach of lower-income (Black) farmers.

Conservation policies and programmes

Erosion of agricultural production has been a perennial issue, and counter measures have cost the state countless billions of rands.

Numerous state-sponsored inquiries into conservation problems have been held. The Drought Investigation Commission of 1923, the Desert Encroachment Committee of 1951, the Commission of Inquiry into Agriculture of 1968 and the Departmental Committee for Veld Deterioration of 1980 provide typical examples. State-sponsored attempts to rectify the degradation of resources have taken a wide variety of forms, including research, education, free advisory services, cheap bank loans, subsidies for conservation works, stock reduction schemes, drought aids, control boards, and laws to control the subdivision, grazing and burning and cultivation of savanna land. It is widely acknowledged that these measures have failed (Bosch 1988). The common response to this observed failure has been a call for bigger and better state machinery with more research, improved advisory services, stricter laws and more active law enforcement. But more centralised state control, guided by government policy rather than by individual pastoralists' needs, is likely to be destructive. Many of the notions pursued and observed in government research, administration and policy have been based on equilibrium models of rangeland function of dubious applicability to event-driven arid and semi-arid systems (Mentis et al. 1989). Elsewhere in the agricultural sector, private research and advisory services have successfully been established despite the subsidised competition afforded by the state (Donovan 1988).

Wildlife conservation

Since the 1960s a private wildlife industry has blossomed in South Africa and is centred around the savannas that provide the habitat for many of Africa's most spectacular animals. The land area involved is difficult to estimate, but far exceeds the area devoted to parks and reserves. Operations vary from pure wildlife (commercial game ranches and private reserves) to mixtures of domestic cattle or sheep with selected antelope species, such as blesbok, impala, kudu and springbok. Few savanna ranches exist without at least some wild antelope present. The contribution to the gross national product exceeds R200 million, derived mostly from live game sales, hunting and tourism. There is a lively trade in live animals, illustrated by the 1990 auction of five black rhinos, purchased by a private concern for R2.2 million.

Several reasons are suggested for the success of the private wildlife industry. First, wild animals in South Africa have the status of *res nullius*. In other words, in their wild state animals are not owned. But the law gives the occupier of land privileges in acquiring or granting ownership (*occupatio*) of the wild animals inhabiting it. Thus with some limitations, the occupier of land may husband and harvest wildlife to his benefit. Given demand for wildlife produce, there is an incentive for the owner, or occupier, of land to conserve his wild animals.

Second, the aesthetic value widely placed on wild animals has promoted wildlife conservation. Indeed, several commercially valuable species such as the blesbok, the black wildebeest and the red hartebeest might have become locally extinct had it not been for the interest of private game ranchers.

Third, commercial hunting has become an increasingly important source of income. Political unrest and economic decline within the traditional wildlife meccas of central and east Africa have forced much of the international trophy-hunting business to move south to Zimbabwe, Namibia, Botswana and South Africa. Increased tourism has generally followed, and is currently estimated to be worth some R4 billion annually to South Africa (including tourism to national parks). Ranchers sell a variety of hunting leases from one or two days to long-term leases. Under the latter, there is incentive for lessees to take proprietory interest in the welfare of the wildlife.

Fourth, with the growing occupancy of the national parks, wealthy businessmen have purchased their own private reserves. Currently the private reserves are highly profitable and cater for upmarket foreign tourists. This has led to suggestions that the state parks should be privatised so that the government does not have to finance their maintenance from consolidated revenue.

State-run parks

In contrast to the private game ranches, several aspects of state-run park management may be criticised. First, historically state-run parks have created little employment for the large, unskilled labour-force that surrounds them. Moreover, the multiplier effects from state-sponsored tourism have not resulted in general social improvement.

Second, state-run parks represent "unfair" competition for the private parks, in that they receive tax-funded subsidies. The conventional argument in favour of state-run parks is that the public has an inalienable right to see the wildlife of their country, and therefore the state must run parks so that all people can visit them. But in reality, only the elite can afford to visit state-run parks, despite the fact that the parks are subsidised, supposedly to keep their prices low. In any case, until recently in South Africa state-run game parks were "Whites only" by law.

Third, the international ban on trade in ivory and rhino horn reduces incentive for wildlife husbandry. The futility of "control by guns alone" was propounded six decades ago (Leopold 1933). Not surprisingly, such bans in east Africa have failed to halt the decline in elephant and rhino numbers. Such restriction of the supply of commodities, like ivory and horn, usually increases their black market price, thus making poaching potentially even more lucrative and therefore attractive.

Fourth, the traditional approach of close management of state-run parks is responding only slowly to contemporary ecological and economic theory. For example, in the Kruger National Park, numerous artificial watering-points have been developed. As explained above, this has relaxed the natural constraints on herbivores, thus increasing their numbers and adversely affecting veld and soil. Subsequently, the culling of animals has been instigated in order to make the herd "fit" the range. Furthermore, in the case of outbreaks of livestock diseases like anthrax, the authorities have immunised individual animals. This conflicts with the basic rationale of a park: to keep things "natural". More cogently, the practices are in flagrant disregard of Holling's (1973) concepts of resilience. If park populations are to be artificially supported, the short-term need for culling (and the costs of management) must rise. In the long term, the self-sustainability of the natural system may be prejudiced.

Overall, the South African wildlife industry, although flourishing compared to that of other African countries, is plagued by bureaucratic vested interests, which interfere with efficient resource allocation and may threaten the sustainable use of wildlife. It is this philosophy, and the practice of nationalisation and central control of wildlife, that have persisted in the African states to the north, as legacies of the colonial era. It is under these circumstances that elephants have been decimated and rhinos have been brought to the brink of extinction.

OPPORTUNITIES TO IMPROVE LAND USE AND INVESTMENT

The present circumstances in South Africa are conducive to redressing many of the ills that have afflicted the country. Specifically in relation to rural land use, the aims might be as follows:

(1) Establish equal access to resources for all races;

(2)   Eliminate provisions that subsidise resource degradation; and

(3)   Restore incentives for productive and sustainable use of resources.

Following are the sorts of provisions likely to be considered in trying to achieve these aims.

Abolition of discriminatory legislation in respect of property rights

The present government has repealed the Group Areas and Land Acts. This will relax past and present constraints on where people live, and will probably promote urbanisation. Urbanisation itself will be traumatic, but it may relieve the extreme pressure on rural-land resources. In the short term, however, free labour movement coupled with economic growth and a rise in urban and industrial wages will, according to the theory of farm-household economics, reduce subsistence activities in the rural areas (Low 1986). In the long term, urban–rural household links may be severed so that urban and rural families operate independently, hopefully leading to rural land becoming more productive than it is now.

The proposal to develop, create or protect property rights means different things to different people. Some see it as a means of protecting White-owned property and preventing Blacks from acquiring land. But that is not the perception of many White farmers, who fear that the repeal of the Land Acts alone (the restoration of equal property rights to all races, thus permitting Blacks to buy Whites' land) will cause them to lose their property. They believe the restoration of equal rights to land ownership would effect land redistribution from Whites to Blacks.

There is, however, a strong call for coercive redistribution of land to compensate for the injustices carried out under the Land Acts. Unfortunately, this form of land reform would not redress past injustices but rather repeat them. Coercive redistribution would require wide government powers of expropriation and land-use control. This paper outlines just a few of the economic problems caused by such central control over property, particularly where resource conservation is concerned. Worse, such a policy would leave the intended beneficiaries of redistribution with no meaningful rights to private property. That is, they would have no protection from further expropriation of their land under future regimes, and very little freedom to use their land as they see fit. The consequences might be as follows. Under expropriation *without* compensation, land values would crash, leaving landowners limited security against which to borrow to effect improvements and to produce. If there was expropriation *with* compensation, then there is the problem of how to fund the compensation. Fixed and movable assets are worth to the order of R100 billion. Raising taxes to meet this is not feasible. Printing money – currency devaluation, as has been done in the past – would depress real interest rates and fuel inflation.

Justice would be served by a system of full rights to private property for all, which would enable individuals or groups to contract freely without government intervention. This would require far more than the repeal of the Land Acts and other racially discriminatory property legislation. It would require also either the repeal of all regulations, tax and credit schemes, subsidies and land laws that protect current landowners from competition for their land, or, alternatively, the extension of these programmes so that they are available to all South Africans. Of these two options the repeal of all existing schemes would be more equitable and would encourage more sustainable forms of land use to emerge. The second option has been tried in the past and failed. Given proper constitutional protection for private-property rights, neither current nor future landowners would receive legal protection from competition for their property. Without such protection, there would indeed be a competitive land and housing market in which Blacks could participate, without fear of future dispossession.

Withdrawal of subsidies

While various subsidy schemes are still in force, the tide of opinion among successive ministers of agriculture in recent years has turned against the payment of subsidies to farmers. Aside from the recognised counter-productivity of subsidies, growing government opposition has arisen from the reduced strength of the rural vote due to urbanisation and from the need to curb rampant government expenditure.

It is likely that the owners of developed farms will, in future, have to fund their own research and advisory services. But the owners of undeveloped land are still likely to be eligible for government subsidies, particularly in terms of access to research results. Currently, the Department of Agriculture is in the process of creating a new, central Agricultural Research Council that will maintain a comprehensive databank covering every farming unit in the country. Unfortunately, the council appears to be little different from the current agricultural information bureaucracy, only with a new command structure. Instead of further centralising the current bureaucracy, government research services (if not fully privatised) should at least be confined to a lean cadre of competent officials who act as a clearing house, hiring private consultants to solve land-users' questions.

Access to capital

Continued "easy credit" is not in the interests of productive and sustainable resource management. It drives up land prices, increases the wealth of existing landholders and rewards poor management. Nevertheless, access to capital is necessary and should be available on terms acceptable to borrowers and lenders, without government intervention.

Land tax

Increasing consideration is being given to the proposal of a tax on the unimproved value of land throughout South Africa, to replace all forms of income tax, with the intent of creating a more equal distribution of land between Blacks and Whites, and encouraging productive land use. The proposal is contentious, with strong arguments for and against. Here, each will be dealt with in turn.

The arguments in favour of a tax on the unimproved value of land include the following:

(1) A land tax would lower land prices and hence supposedly make it easier for Blacks to acquire land. Moreover, as potential rises in land value would be taxed away, there would be less incentive to hold land as a means to maintain and increase wealth.

(2) Present taxation is on labour (effort), which acts as a disincentive to labour and encourages people to substitute land for labour.

(3) Landowners would be induced by a land tax to maintain their resource base, lest the owner no longer be able to pay the tax or sell his land at a reasonable price. Under the present system, if the resource base declines, the land becomes less productive and less tax is paid.

(4) Taxing land is a potential means of countering the reported declines in small-farm production that have arisen despite improved technology and rural development programmes (Low 1986). Taxing the land would be an inducement to make the land more productive.

(5) A land tax, or, more accurately, taxation of the privilege of the right of access to land, might be more compatible with African and conservationist values than is the classical liberal notion of individual property rights. In terms of the concept of taxation of the privilege, provided the tax is paid the landholder retains usufruct. The privilege is exclusive and can be exchanged voluntarily, and there is an inherent inducement to respect land which is, as it were, "borrowed" from posterity.

200

The arguments against the land tax include the following:

(1) A tax on the unimproved value of land will not help the poor. The land buyer will be taxed on how much he or she is expected to gain from developing it – so developing, or "improving" the land will be more expensive than it would without the tax. Conversely, the buyer cannot earn as much revenue from developing the land as he or she would without tax. That will, indeed, work to reduce the price; buyers will not be willing to pay as much for the land, due to the heavy costs of improving it. But if a poor person then buys the land at the reduced price, he or she will face the same heavy costs of developing it (or the same low rate of return on the land). There won't be any point in the poor buying land, since they will not be able to afford to build on it, because of the tax.

Supporters of the land tax argue that people will still buy such land and they *will* improve it – in fact, the tax will force them to use the land extremely productively, in order to pay the cost of the tax *and* earn a profit. In other words, landowners will be motivated to use the land well, its supporters say. But only the wealthy and highly skilled will have the resources to do so – to make the land "extra productive". The low-income family that wants to buy a small plot of land on which to live or start a small business will find it prohibitively costly.

The most likely result of all this is that the very rich will buy and develop land, while the poor and middle-income earners will not be able to afford to. Instead of land being redistributed throughout the population and used efficiently, much of it will likely remain unused while the poor remain landless.

(2) It is a lack of taxation that induces individuals to be productive. While some level of taxation may be necessary, it is still a disincentive to produce; it prevents individuals from keeping all the benefits of their effort and employment of resources. A land tax will only increase production if the total amount of tax collected from each individual is lower under a land-taxation system than under any other kind of taxation. A land tax small enough to meet that requirement would pose no threat to current owners of large and inefficient tracts of land. That is, the tax would not be large enough to induce those landowners to sell off pieces of their land to small-scale farmers, as the land tax is intended to do.

(3) Several methods of determining the level of land taxation are being proposed. One is the nomination of taxable land value by the owner himself, so long as the state reserves the right to expropriate that land at, say, 120 per cent of the nominated value (to prevent undervaluation).

But these expropriation powers would be open to abuse by vested interests. That is, the state would have wide and unchecked powers to expropriate individuals' property in order to give or sell it to others with greater political influence.

(4) It is not at all clear that a land taxation system is more compatible with African custom than classical liberal property rights. Under African customary law, individuals are not permitted to sell land, but there is a strong system of individual land-use and ownership rights. At best, one can say that African customary law shares common themes with both the Georgeist philosophy, from which the land tax is derived, and classical liberal philosophy, on which a free land market is based.

Since previous methods of resource conservation enforcement by government has failed, the land tax appeals to many simply because it is new and untried. The present authors disagree with each other on the proposed merits of a land tax but, as the tax is being seriously considered as part of a post-apartheid land-reform package, it warrants consideration and further debate.

CONCLUDING REMARKS

The most constructive role the South African government could play in the promotion of better savanna land use and conservation would be the removal of legal restrictions on the land market.

The current system of strong central control over land use has been an object lesson in unintended consequences. Government intervention intended to promote conservation has had the opposite effect. Central economic planning, as in strict control over agricultural markets, has prevented landowners from conserving land properly and reduced opportunities to increase production sustainably. Apartheid has created a great disparity between the White urban rich and the Black rural poor, preventing most Blacks from owning property and promoting migrant labour. In terms of the farm-household economics theory, this has induced Black rural households to purchase goods from urban areas rather than produce them locally. Development programmes have failed to integrate rural production with the urban and industrial labour markets. Rural Black areas have become net importers of goods.

Together these policies have sheltered inefficient management practices from the competitive pressures, albeit harsh, that promote productive and sustainable land use. A free, or at least far freer, land market, where land ownership and use is regulated by economic competition without

202

government assistance to any of the parties involved, would restore the incentives for effective conservation that South Africa desperately lacks.

In contrast to our proposals of non-interventionist government policy, there is a call in South Africa for concerted government intervention in the land market to effect post-apartheid restitution. However well-intended such arguments are, we believe them to be misguided. The South African case is a seminal lesson in the failure of interventionist land-use policy. To us it is clear that ultimately most interventionist forms of government harm even those they are intended to help. In the case of South Africa's savannas it would leave the victims of apartheid worse off.

There is grave danger in South Africa that the mistakes of the past will be repeated, this time in the name of restitution for the poor and disadvantaged. Real restitution requires the development of a market economy in which South Africans enjoy full private-property rights for the first time in three centuries.

REFERENCES

Bauer, C. (1983) A tragedy of the commons? *Free Market Magazine* 3 (1):18–22.

Bosch, J.O.H. (1988) Vegetation deterioration in southern Africa: A research and researcher problem. *Journal of the Grassland Society of Southern Africa* 5:34.

Bundy, C. (1988) *The rise and fall of the South African peasantry*, 2nd ed. John Philip, Johannesburg.

Caldwell, D. (1989) *South Africa: The new revolution*. Free Market Foundation, Johannesburg.

Caughley, G. (1976) Wildlife management and the dynamics of ungulate populations. In: T.H. Coaker (ed.) *Applied biology*. Academic Press, London.

Clark, C.W. (1985) *Bioeconomic modelling and fisheries management*. Wiley and Sons, New York.

Danckwerts, J.E. (1981) A technique to assess the grazing capacity of sweetveld with particular reference to the false thornveld areas of the Ciskei. MSc thesis, University of Natal, Natal.

Danckwerts, J.E. and King, P.G. (1984) Conservative stocking or maximum profit: A grazing management dilemma? *Journal of the Grassland Society of Southern Africa* 1(4):25–8.

Donovan, P.A. (1988) Management of research and development at the sugar industry's Experiment Station at Mount Edgecombe, Natal. *South African Journal of Science* 84: 793–6.

Edwards, P.J. (1981) Grazing management. In: N.M. Tainton (ed.) *Veld and pasture management in SA*. Shuter and Shooter in association with University of Natal Press, Pietermaritzburg.

Hardin, G. (1968) The tragedy of the commons. *Science* 162:1243–48.

Holling, C.S. (1973) Resilience and stability of ecological systems. *Annual Review of Ecology and Systematics* 4:1–23.

Jones, R.J. and Sandland, R.L. (1974) The relation between animal gain and stocking rate; derivation of the relation from the results of grazing trials. *Journal of Agricultural Science* 83:335–42.

Koyana, D. (1980) *Customary law in a changing society*. Juta, Johannesburg.

Leopold, A. (1933) *Game management*. Charles Scribner's Sons, New York.

Low, A. (1986) *Agricultural development in Southern Africa: Farm-household economics and the food-crisis*. Heinemann, Portsmouth.

Mentis, M.T. (1981) Amatola Basin Rural Development Project (Ciskei) II: The grazing and browsing capacity of the Amatola Basin. Agricultural and Rural Development Research Institute Report 44/81, Fort Hare.

Mentis, M.T. (1984) Optimising stocking rate under commercial and subsistence pastoralism. *Journal of the Grassland Society of Southern Africa* 5(1):20–24.

Mentis, M.T. (1986) Range dynamics by classical succession and strategic modelling. In: P.J. Joss, P.W. Lynch and O.B. Williams (eds) *Rangelands: A resource under seige.* Australian Academy of Science, Canberra, pp. 19–21.

Mentis, M.T. (1990) Are multi-paddock grazing systems economically justifiable? *Journal of the Grassland Society of Southern Africa* 8 (1):29–34.

Mentis, M.T., Grossman, D., Hardy, M.B., O'Connor, T.G. and O'Reagain, P. (1989) Paradigm shifts in South African range science, management and administration. *South African Journal of Science* 85:684–7.

Organisation of Livestock Producers (1988) Investigation into the red meat price forming process. Submission to Committee of Investigation into the Red Meat Price, Vryburg.

Tapson, D. (1990) A socio-economic analysis of small-holder cattle producers in KwaZulu. PhD thesis, submitted. Vista University, Republic of South Africa.

CHAPTER 10

DEVELOPING BOTSWANA'S SAVANNAS

D.W. Pearce

SUMMARY

In Botswana's savannas, as in many other countries, subsidies, taxation arrangements and price-support policies tend to encourage degradation. The Lomé Convention, which gives certain Botswana beef producers preferential access to European Community markets, aggrevates this process and narrows the production base. Under a Tribal Grazing Lands Policy, significant parts of the nation's savannas have been converted to private ranches.

The dual land-tenure system mixes private and common-property regimes in a manner that creates a non-caring attitude to savanna management and, inequitably, is placing additional pressure on communal grazing land. Wealthy pastoralists have been given access to private ranching areas and encouraged to develop new watering points but have been caught by a *mafisa* system that obliges "ranchers" to allow cattle owned by family and kinsmen onto "their" land. They, however, also take advantage of the *mafisa* system and continue to graze their livestock on common-property land. Existing administrative arrangements have transferred management from the community to land boards without supplying them with adequate legal and regulatory resources.

Solutions to these problems lie with the removal of environmentally destabilising incentives and the separation of property rights so that people have a clear right to utilise community grazing areas or private ranches but not both. There is also a case for the careful development of tourism in some parts of the savannas so that wildlife and livestock management become more complimentary.

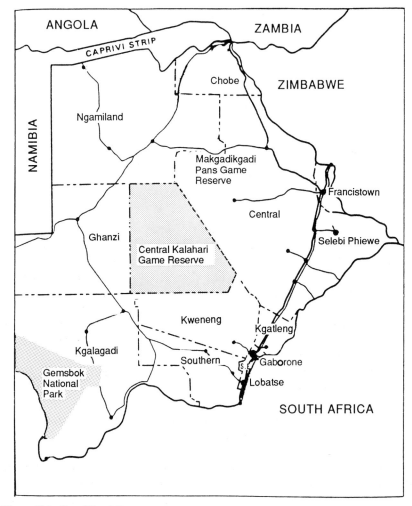

Figure 10.1   Republic of Botswana

INTRODUCTION

Botswana is a landlocked country with borders shared with Zimbabwe, South Africa, Namibia and Zambia (Figure 10.1). The land area is approximately 580,000 sq km, but most of its population of 1.2 million live in the eastern margin drained by the Limpopo River. In the north-west is the Okavango delta, an inland swamp arising from the River Okavango which drains inland from Angola; this swamp is one of the most remarkable wildlife areas anywhere in the world. The remaining two-thirds of Botswana

206

Table 10.1 Botswana: Population (000s). Source: Central Statistical Office

	1971	*1981*	*1991*
Urban	63.5	166.3	341.1
Rural	510.5	774.8	1,006.4
Total	574.1	941.0	1,347.6

is occupied by the sands of the Kalahari Desert where surface water is present only after the rains, and then only in "pans" (shallow depressions). Rainfall is variable and generally comes in short, intensive bursts; from 1982 to 1986 there was a major drought.

By African standards, Botswana is relatively wealthy. On the World Bank classification it ranks as a lower middle-income country with a per capita income in 1986 of US$ 840, the equivalent of that of Jamaica. The major part of this economic success is owed to the mining of diamonds, copper-nickel and coal. Botswana also has a significant livestock sector based on cattle. The environmental challenges in Botswana are several: the overgrazing of cattle on fragile soils has led to land degradation; cattle conflict with wildlife in terms of competition for land and through the effects of veterinary fences on migratory animals; and accessible water is scarce, causing pressure on groundwater and a focus on the Okavango delta as a potential source for diverted water. Although rich in wildlife, including the inhospitable Kalahari regions, Botswana has deliberately not pursued a policy of developing its wildlife resource. Tourism is moderate in comparison to the diversity and stock of wildlife and in comparison to the generally well-developed communication system.

The 1981 census indicated the population data shown in Table 10.1. The total population of under 1 million people is broadly split, 80 per cent in the rural areas and 20 per cent in urban areas, but this is expected to change to 75 and 25 per cent respectively in the early 1990s. Population growth, in general, is fast at 3.6 per cent per annum.

The ethnic composition of the population is difficult to determine accurately. The number of "bushmen" is unknown, these being the Basarwa (San) people, who first inhabited what is now Botswana and could comprise anything from 2 to 25 per cent of the population (Cooke 1988), with very few of them still occupying the lifestyle made so famous by van der Post (1958), most being assimilated in other communities. The Bamangwato account for about 40 per cent of the population, being one of the cattle-herding Tswana peoples who invaded Botswana from Zimbabwe and South Africa about 300 years ago. Other Tswana peoples are the Bakwena and the Bangwaketse.

Table 10.2 Water demand in Botswana, 1987 ($\times 10^6$ cubic metres per annum). Source: Background papers to draft Botswana National Conservation Strategy

Urban	21
Major villages	5
Rural villages	2.5–4.5
Mines and energy	17
Livestock	45–50
Irrigation	30–40
Total	120.5–137.5

In 1891 the British Cape Colony Protectorate was extended to Botswana and tribal lands were assigned to eight recognised Tswana tribes, three of them (the Bamangwato, the Bakwena and the Bangwaketse) occupying the best agricultural land in what are now Central District, Kweneng District and Southern District respectively. Still other tribes moved in from the south and some from the north, producing a rich ethnic variety of population which, none the less, has created political tension. In general, the Tswana elite have remained in political control of Botswana and various efforts have been, and are being, made to distribute national benefits in a more widespread manner. The ethnic composition, and particularly the Tswana dependence on cattle, is important for an understanding of the relationship between the Botswana environment and the economy.

THE STATE OF THE ENVIRONMENT

Water

Water is a major constraint on Botswana's development. The 5 years after 1981 were characterised by below-average rainfall. The major users of water are currently irrigation and livestock, about 35 per cent of total consumption each, with mining and urban demand accounting for perhaps a further 12 per cent each. Total demand was some 120–140 cubic metres in 1987. An approximate breakdown of demand is given in Table 10.2.

Growth in demand is expected to be substantial. Total domestic demand is expected to increase by a factor of 5 by 2007, while demand for livestock and mining uses will grow only slowly. The other sector that might grow dramatically is irrigation depending on national policies towards food security. Some estimates suggest a tenfold increase in demand by 2007, making irrigation responsible for 60 per cent of projected demand. But how far a policy of expanded irrigation is feasible is uncertain.

How far existing consumption rates exceed the sustainable yield of water from rivers and groundwater, reached by boreholes, is not known. But the major growth in demand for water has been in the south-east where water resources, as far as they are known, are in potential deficit. There is a regional imbalance between supply and demand not only for geographical reasons – highest demand occurring where surface water is in short supply – but also because some catchment areas lie outside Botswana. The implications are either that groundwater sources, the extent of which is not known accurately, need to be developed or that some form of surface water diversion is necessary. Groundwater sources primarily supply the livestock industry and urban sectors. Surface water primarily supplies irrigation and wildlife.

Water shortages have shown up most in terms of precluded development – i.e. land developments have not occurred because of the absence of identified water supply. The livestock sector has also suffered because of the difficulty of finding new boreholes[1]. Urban supply has been rationed at times of rainfall deficit.

Land degradation

Livestock numbers have grown rapidly in Botswana. Figure 10.2 shows that the growth has generally been related to population change, with some more erratic variation in numbers of smallstock (e.g. goats). Overgrazing, in the sense of animal numbers in excess of the range's carrying capacity, appears to be common. Arntzen and Veenedaal (1986) assembles available data to show that overstocking occurs throughout Botswana. The main holders of cattle are the relatively wealthier households: 8 per cent of farmers held 45 per cent of cattle stock in 1984; and 30 per cent of cattle holders held 4 per cent of the stock (Arntzen and Veenendaal 1986). There is little systematic investigation of the cattle-range quality linkage, but most commentators agree that much of range degradation arises from excess livestock numbers (Cooke 1985; Ringrose 1986). Thus bush encroachment – the loss of palatable perennial grasses in favour of less palatable species and woody biomass – appears to emanate from centres of cattle concentration. Soil erosion can ensue as the protective functions of the grassy cover are lost, although evidence on soil erosion in Botswana is very limited.

Interpretation of Landsat imagery by Ringrose (quoted in Arntzen and Veenendaal 1986) suggests fairly extensive land degradation. She finds

---

[1]An aquifer can serve several different boreholes, so that simply sinking extra boreholes in a given area is not a solution to the scarcity problems. Indeed the issue is exacerbated by conflict between users. In Botswana boreholes must be at least 8 km apart, to reflect this fact, and to maintain the surrounding grazing area.

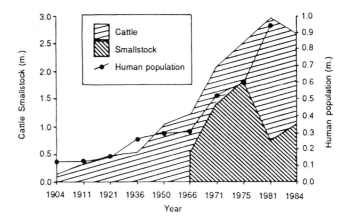

Figure 10.2 Cattle and smallstock numbers in Botswana (× 1000) Source: Arntzen and Veenendaal (1986)

severe degradation around some villages, degradation in western Botswana and around the Okavango delta, and elsewhere. However, the relative contributions of climate and man-made factors appears to be unknown. Time-series analysis is generally not available, but, in a further paper, Ringrose (1986) found marked increases in the extent of exposed soils in the area round Gaborone in south-east Botswana between 1982 and 1984. The changes were particularly notable in the communal grazing areas close to Gaborone and in the settled areas of the eastern Kalahari. Moreover, actively growing vegetation declined in the Kalahari and agricultural lands to the east.

There is some evidence that crop production is contributing to soil erosion in some areas, but again adequate time-series analysis of the possible linkages between choice of crops and soil quality is lacking.

Loss of woody biomass

There is no systematic inventory available to determine wood stocks in Botswana. One study (Environmental Resources Ltd 1985) of eight regions in eastern Botswana suggested that fuelwood supply was only just keeping pace with demand in the south-east region and that demand would outstrip supply by the end of the 1980s. Elsewhere few regions faced prospective risk of major shortage, but some of the major towns other than Gaborone (the capital of Botswana) already experience some shortages. In fact the situation is worse than is revealed by an analysis of fuelwood demand alone, since other wood demands (e.g. poles for building) can more than double demand based on fuelwood needs only. The study suggested that fuelwood demand would increasingly involve a switch to less preferred species, a rise in fuelwood prices, spontaneous conservation measures (reduced cooking, use of animal dung for fuel) and, for the richer households, a switch to petroleum products.

Wildlife

Botswana is immensely rich in wildlife. The 1981–1986 drought affected populations significantly, but extensive numbers of the larger animals – buffalo, antelope, elephant, hippopotamus, giraffe, etc. – remain, and Botswana has a virtually unparalleled population of birds and smaller wildlife. Controversy surrounds the extent to which cattle fences have interfered with migration patterns and contributed to mortality. Williamson (1987) has analysed the effects of the Kuke fence, built in 1958 and running west-east to the south of the Okavango delta. At the time of the fence construction, migrating wildebeest died because the fence blocked their migration path, or forced them into the Lake Xau area in large numbers where many died. Williamson estimates the deaths at "hundreds of thousands of wildebeest alone, plus nobody knows how many deaths of other species". Others have disputed the extent of the losses.

The main impression left from this brief review of Botswana's environment is a familiar one: lack of information makes it difficult to assess both the existing stock of natural capital and its rate of change over time. There is sufficient evidence, however, to suggest that Botswana's major problem is a threatened water shortage. A wood supply problem seems of less concern. Range degradation does appear to be occurring and, regardless of the relative contributions of overstocking and climatic variability to the problem, it seems clear that some form of improved range management is required. For wildlife, the problem lies with the direct competition with cattle for land, an issue that is likely to be best resolved by a more positive approach to the sustainable management of wildlife for economic benefit. The rest of this chapter develops these themes.

The economic contribution of natural resources

Table 10.3 shows the relative contributions of different sectors to the gross domestic product (GDP) of Botswana. The sectoral composition of GDP indicates Botswana's high level of dependence on natural resource-based activities. In 1984–1985, the beginning of the current plan period, mining

Table 10.3 Real gross domestic product (GDP), by sector: 1980–81 to 1985–86 (at constant 1979–80 prices in million pula). The 1984–1985 and 1985–1986 figures are preliminary, the latter determined by industrial activities only. The dummy sector is a correction for imputed bank service charges. In 1985–1961 pula = approx. 0.53 US dollars. Source: Botswana CSO Statistical Bulletin 1987

	1980–81	*1981–82*	*1982–83*	*1983–84*	*1984–85*	*1985–86*
Agriculture	75.0	71.8	60.1	50.9	48.0	48.0
Mining and quarrying	260.6	222.1	393.3	533.4	560.6	673.2
Manufacturing	37.0	45.8	42.4	44.0	46.0	45.0
Electricity and water	15.3	15.9	15.7	19.5	19.6	20.7
Construction	32.0	37.2	26.2	38.7	32.3	30.3
Wholesale and retail trade	163.8	150.7	162.1	182.2	204.4	223.4
Transport and communications	14.8	18.1	23.8	23.1	25.2	26.1
Financial institutions	49.7	55.6	58.9	65.2	67.9	74.0
General government	104.7	113.9	122.9	136.5	156.2	165.7
Household, social and commercial services	25.6	29.2	35.0	34.9	34.9	35.5
Dummy sector	−16.7	−17.0	−19.9	−22.4	−24.2	−26.3
GDP	761.8	743.3	920.5	1106.0	1170.9	1315.6

and agriculture together accounted for 54.5 per cent of GDP, 92.3 per cent of exports and approximately 90 per cent of all employment, formal and informal. Significantly, however, within the natural resources area there are extraordinary differences in the performance of the exhaustible and renewable resource-based sectors. Where real mining GDP grew at an average annual rate of 21 per cent between 1980–1981 and 1984–1985, real agricultural GDP fell over the same period at an annual average rate of just under 11 per cent.

The high level of dependence on natural resources is sufficient to justify a resource conservation policy. Moreover, the dominance of mining is not expected to continue, which means that the balance of the economy has to switch some form of manufacturing or expansion of services or of agriculture, or some combination of these. This diversification of the economy is consistent with the principle of sustainable development since reliance on a single activity subjects the economy to exogenous risks (e.g. world price variations). This, in turn, suggests that agriculture has to be rehabilitated and that other resource-based activities need to be developed.

Table 10.4 shows some crude estimates of the contribution of various natural resources to the economy of Botswana. The table suggests that activities not typically accounted for in the GDP estimates might be worth 100 million pula per annum, or perhaps 7–8 per cent of GDP. The potential is clearly significantly greater.

Table 10.4 Economic contributions derived from the natural resources of Botswana. The estimates are derived from a wide variety of sources, and relate solely to the annual flow of benefits – i.e. changes in the capital value of the resource are *not* included. Furthermore, the figures do not all refer to the same year. Thus the figures need to be regarded as *preliminary*, indicative estimates. In addition, the tourism-related craft industries provide at least 3500 jobs. Source: Perrings et al. (1988)

Resource	Gross output (million pula)	Value added (million pula)	Employment (*rounded*)
Rangeland			
Mainly livestock commercial	76.7	54.5	14,300
freehold	17.7	0.4	
Veld products commercial*	4.4	n.a.	2640
subsistence	50.0	n.a.	n.a.
Wildlife/tourism photographic			
safari	13.4		
hunting safari	2.6	2.9	2020
wildlife farming	0.1		
trade/processing	0.9		
subsistence	7.5–11.5	6.2	5020
Fisheries			
Commercial	4.7	n.a.	900
Subsistence	2.6	n.a.	5000–11,000
Forestry			
Commercial	n.a.	n.a.	n.a.
Traditional	10.3	n.a.	n.a.
Wood fuels			
Urban	13.0	n.a.	n.a.
Rural	16.6	n.a.	n.a.

*These include mokola palm, mopane worm, grapple, plane, silk cocoons, thatching grass.

RESOURCE DEGRADATION AND MANAGEMENT POLICY

In the remainder of this chapter we look at the resource sectors in light of the factors giving rise to resource degradation or the risk of degradation, and the policies that might be adopted to combat resource loss.

The mining sector

Over 9000 people are employed in Botswana's mining sector, about 7 per cent of all formal sector employment[2]. Although currently mining

[2]A further 20,000 are recruited for work in South Africa's mines.

contributes approximately 40 per cent of GDP, and half of government revenues, its contribution is expected to decline in the future. The economic issue is how to utilise the "rents" that have been created from this mineral wealth for the development of the rest of the economy along a sustainable path. Botswana has typically pursued a policy of rapid extraction of its mineral wealth, which is justified by the low expectations of future real price increases[3]. Diamond extraction is coupled with a policy of storing the output. This can be justified by the fact that storage costs are low, extraction costs are a small fraction of the diamond price and fixed costs of extraction are a high proportion of total costs. Additionally, continual extraction maintains employment in the industry, and this can be thought of as a "positive externality" by the industry which further justifies the extraction profile.

Any environmental degradation associated with mining would offset this presumption that rapid exploitation is justified. Essentially, any environmental costs should be added to the costs of extraction to determine the real social cost of extraction. The higher the environmental cost, the higher is the real cost of extraction, and this should modify the policy of rapid extraction. There is air pollution from Botswana's mining activities, and emissions at one mine, Selibe-Phikwe, do appear to exceed government standards. There may also be some contamination of groundwater sources used to meet the mining sector's water demands. Overall, however, it seems unlikely that resource conservation objectives should alter the pattern of exploitation of Botswana's minerals unless clearer evidence of associated environmental degradation comes to light.

Livestock

Botswana's livestock increased dramatically up to the 1980s, receiving a setback only between 1982 and 1985 when drought reduced the cattle herd from approximately 3 million to 2.5 million. Cattle have a high cultural importance in Botswana. Briefly, to be a "real Motswana" (an individual of Botswana) one must own some cattle. The extent to which an individual's opinions are respected varies directly with cattle ownership. A complex system exists whereby cattle are loaned between individuals (*mafisa*), partly based on the need to distribute cattle so as to minimise the risks of raiding

---

[3]The rationality of rapid extraction in this context lies with Hotelling's rule for optimal extraction of a finite resource. If expected real price increases are higher than prevailing interest rates, it is better to leave the resource "in the ground" to collect capital appreciation. However, if price expectations are below the interest rate, it pays to extract the resource rapidly and invest the proceeds at the ruling interest rate.

and partly to insure against pests and drought. Thus the larger the herd that is owned, the more secure an owner may feel against the risk of disease and drought, and the more he can assist smaller herd owners by lending cattle under the *mafisa* system to help with reconstituting herds affected by drought. Cattle play a major role in *rites de passage* – that is, they have a ceremonial role at birth, marriage, achieving adulthood, etc. In rural areas there are few alternative opportunities for employment, but many cattle are owned by absentee owners in towns. As the human population expands, so does the cattle population.

Livestock ownership is directly encouraged by government policies which provide fiscal incentives to overstock. Many inputs are subsidised, as with veterinary fences and veterinary services in general. Extension and research services are provided by the government, as are slaughterhouse facilities. A critical incentive is the ability to write off losses in agriculture against income from other activities, so that there is no incentive to manage herds to maximise pre-tax gains. The pricing system has also encouraged overstocking. During the earlier years of the 1981–86 drought, the Botswana Meat Commission chose to pay the highest possible price to sellers to secure short-term gains, encouraging overstocking rather than understocking. Moreover prices tend to be lowest at the onset of the dry season, providing no incentive to sell at that point, despite the fact that it would avoid cattle making demands on the rangeland environment at its most susceptible time. The European Economic Community (EEC) market gives preferential treatment to Botswana beef under the Lomé Convention, further encouraging increased herd size.

Criticism of Botswana in terms of policies encouraging environmental degradation has centred on the livestock sector, and especially the alleged help given to the expansion of the sector by international aid and lending agencies. Early efforts to redistribute the cattle stock included the First Livestock Development Project, approved by the World Bank and co-financed by Sweden in 1972. This aimed to shift the expansion of commercial ranching to the western lands to remove overgrazing pressure in the east. The project failed because of the difficulties of monitoring activity in these remote areas; because what benefits there were did not reach the intended population of the west and the north; and because of the unwillingness of ranchers to pursue the management policies that were an integral part of the project.

In 1975 the government in Botswana introduced the Tribal Grazing Lands Policy. To understand that policy it is necessary to have some grasp of land tenure in Botswana; the three basic types of tenure are illustrated in Figure 10.3. Approximately 32 per cent of land is state land, including reserves and national parks; just 3 per cent is freehold land, privately owned; and 65 per cent is tribal land. Originally, under the Protectorate,

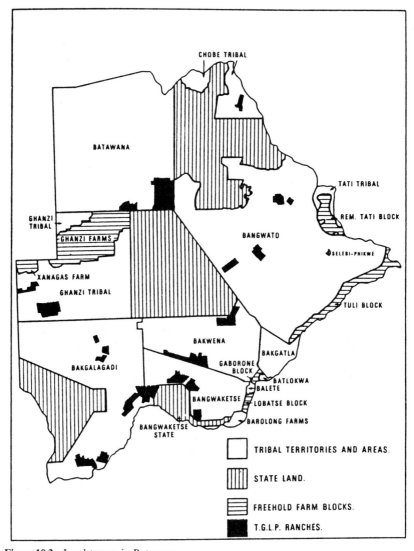

Figure 10.3 Land tenure in Botswana

control of tribal land was exercised by the tribal chiefs. The Tribal Lands Act 1968 transferred control to District Land Boards, the aim being to ensure a better and fairer allocation of land use. The boards were to determine an appropriate land use and allocate use rights to individuals, while the ownership remained communal. At the same time, the development of boreholes established a new kind of *de facto* right in remoter areas – he who drilled the borehole and paid for it controlled the water and hence, effectively, the land within a day's walking distance of it.

The Tribal Grazing Lands Policy (TGLP) was aimed at controlling overgrazing in the tribal lands. It set up fenced leasehold ranches on the lands (Figure 10.3), designed to absorb the large herds. The aim was to leave the rest of the tribal lands available to smaller ranchers who would continue communal grazing, and a further proportion was to be left as "reserve" areas in preparation for wildlife management schemes. Thus tribal lands were to be split as large commercial leasehold ranches, smaller commercial units and tribal communal grazing and potential wildlife management areas. The District Land Boards were to be responsible for deciding land-use plans, now supported by representatives of central ministries in light of doubts about the land boards' effectiveness. But the land boards faced opposition from tribal groups, used to the more traditional allocation of use rights through their own institutions, and distrusting the boards because of some of the decisions already made.

The TGLP also fell foul of the "dual rights" problem. Botswana deciding to go in for the commercial (leasehold) ranching did not give up their traditional tribal rights to the communal land. They could thus degrade their leasehold land and then move their cattle on to common land and wait for the overgrazed land to restore itself. Additionally, through the *mafisa* system, the ranchers would allow other cattle owned by family and kinsmen on to their land, furthering the degradation process.

In 1977 the Second Livestock Development Project was started, aimed at implementing the TGLP. The aim was to encourage the development of the commercial leasehold ranches and to protect the interests of the small farmer left on the communal lands. In the event, the policy failed for several reasons. First, the take-up of TGLP ranches was small and large herd owners still refused to adopt modern management practices. Secondly, the dual-rights issue was not tackled. Thirdly, while small farmers were encouraged to move to allegedly virgin lands occupied by wildlife and, perhaps, by bushmen, less of these lands were suitable for cattle than was thought hitherto. In addition, development costs were much higher than expected.

The debate over the appropriate policies to pursue continued into the 1980s. The World Bank focussed attention on the fiscal incentives to overgrazing, while the Botswana government tended to stress measures to protect the communal grazing areas, including the modification or abolition of dual rights. In 1985 a Land Management and Livestock Project was launched. This focussed on the preparation of land-use plans consistent with the carrying capacity of the rangelands, gives credits for ranch development, has funding for livestock extension services and seeks to strengthen central institutions' planning capacity. It strengthens the land boards. It also called for a price incentive study, which was eventually released in 1987 (McGowan International 1987). But, as the report itself notes, its terms of reference were changed by the government to exclude discussion of taxation allowances. At the time of writing, this aspect of the work still has to be completed.

The livestock sector thus remains controversial, especially so because the alleged environmental costs of international aid to Botswana were highlighted in a publication, *Bankrolling Disasters*, produced by environmental organisations in 1986. The essence of the criticism was that commercial ranching is not suited to the tribal lands, that wildlife was being "squeezed out" by the expansion of cattle lands, wildlife migrations were impaired by fences, and the benefits of the livestock development projects accrued mainly to the wealthy. While some of the criticisms rest on fairly extensive misinformation about the nature of the component parts of the development projects, the general focus on overgrazing is actually shared by all parties. The dispute is about what to do. It seems clear that only a policy that effectively addresses the use and land-tenure rights of cattle owners, and that deals with the incentives to overstock, can succeed.

Undoubtedly, there is conflict between wildlife and cattle and this raises the issue of the extent to which wildlife management can substitute for cattle ranching. Various estimates exist to show that some forms of wildlife ranching can yield economic rates of return above those achieved by cattle ranching, but assuming no fiscal incentives for cattle ranching such as exist at the moment. That, together with other rural development concepts, defines the proposed Botswana National Conservation Strategy, sponsored by the International Union for the Conservation of Nature and still being completed at the time of writing (early 1989).

Arable farming

Perrings et al. (1988) note the major difference in productivity between arable farms. Traditional farms tend to be on communal land, whereas commercial farms are on leasehold or freehold land. The land tenure argument previously outlined would suggest that we might expect output per hectare to be higher on commercial than on traditional farms, and this is indeed the case. However, incentives to conserve land in which there are distinct rights are only one factor in a complex of reasons for this differential productivity. Commercial land also tends to be located in areas of better rainfall. Virtually all irrigated land is freehold. Moreover, poor farmers tend to engage in risk-averse strategies so that output tends to be maintained at low but reliable levels compared to the higher output levels obtainable with a more "adventurous" profit-maximising strategy. The differential productivity also reflects the lack of more advanced technologies in communal areas, where "advanced" can include the use of animal draught power.

TOWARDS SUSTAINABLE DEVELOPMENT

How, then, does this brief overview of issues assist in identifying a sustainable development path for Botswana? A number of issues have emerged, as follows:

(1) Future economic development in Botswana needs to be more broadly based. Progress in the past has been largely in the mining sector, with the agricultural sector being allowed to decline. Past development has also paid little or no attention to the economic potential of the rich wildlife and natural habitat resources of Botswana.

(2) The existing structure of price and income incentives in the agricultural sectors – arable and livestock – tends to encourage resource degradation. Such incentives include tax write-offs and subsidies to livestock and the existing arrangement with the EEC with respect to elevated beef prices. Resource conservation considerations suggest a lowering of these administered prices and a reduction in the subsidies and tax allowances. Grain prices on the other hand, tend to be linked to subsidised prices in South Africa, making for disincentives to produce grain. Farm input subsidies (e.g. in the form of land clearance allowances) have probably made environmental conditions worse.

(3) The existing structure of land tenure acts as a "disabling" incentive. Overgrazing is encouraged in communal lands, and the existence of dual-grazing rights, whereby landowners can still graze communal land, both encourages a non-caring attitude to the commercial land and adds to the pressure on communal land.

(4) Laws and regulatory capacity in place are not used. The land boards in Botswana have the power to control the allocation of grazing lands, arable land and water rights. Lack of progress in securing an allocation system that conserves the natural capital of the land is partly explained by the sheer political difficulty of modifying property rights in a country where cattle have more than economic significance. But if *communities* could be given rights of exclusion over defined areas of grazing land, then the principle of communal tenure can be preserved while avoiding the worst excesses of open-access resource use. Indeed, the aim should be to establish common property in its true sense, as a form of defined communal ownership, and prevent *de facto* open-access use whereby anyone can lay claim to grazing rights. Such a

policy would have to include the abolition of dual grazing rights. Some evidence of a shift of attitude, in this respect, already exists, as with exclusive assignment of water rights to syndicates of farmers.

(5)   Underutilised resources should be "developed" so as to be both conserved and managed for a sustainable yield. This suggests the development of some of the rural products, especially for tourism.

Overall, a judicious mixture of removing environmental destabilising incentives, introducing enabling incentives based on resource rights and focussing modest financial resources on the development of certain rural industries would assist in the sustainability of Botswana's economy.

REFERENCES

Abel, N., Flint, M., Hunter, N., Chandler, D. and Maka, C. (1987) *Cattle keeping, ecological change and communal management in Ngwakerse.* Integrated Farming Pilot Project, Gaborone.

Arntzen, J. and Veenendaal, E. M. (1986) *A profile of environment and development in Botswana.* Institute for Environmental Studies, Free University of Amsterdam, Amsterdam.

Cooke, C. (1988) Botswana land management and livestock: A case study. Presented at Africa Region Environmental Seminar, Annapolis, Maryland, 18–20 May.

Cooke, H. J. (1985) The Kalahari today: A case of conflict over resource use. *Geographical Journal* 151.

Environmental Resources Ltd (1985) *A study of energy utilisation and requirements the rural sector of Botswana.* Overseas Development Administration, London.

McGowan International (1987) *National land management and livestock projects: incentives/ disincentives study.* McGowan International, Gaborone.

Perrings, C., Pearce, D.W., Opschoor, H., Arntzen, J. and Gilbert, A. (1988) *Economics for sustainable development: Botswana – a case study.* Ministry of Finance and Development Planning and Ministry of Local Government and Lands, Gaborone.

Ringrose, S. (1986) Desertification in Botswana: Progress towards a viable monitoring system. *Desertification Control Bulletin*, No. 13.

van der Post, L. (1958) *The lost world of the Kalahari.* Penguin, Harmondsworth.

Williamson, D. (1987) *An environmental analysis of Botswana.* IUCN, Conservation Monitoring Centre, Cambridge.

CHAPTER 11

RESTORING THE PRODUCTIVITY OF INDIAN SAVANNAS

M. Gadgil

SUMMARY

Savannas secondarily derived from woodland ecosystems cover a large area of India. They perform an important service in meeting a variety of subsistence biomass needs of a significant proportion of India's rural population. Largely controlled by the State, the savannas are currently being degraded under the pressure of unregulated harvests by the commercial as well as the subsistence sector.

These undisciplined harvests are promoted by the open-access regime prevalent on State-controlled lands and by the high level of subsidies for commercial forestry. This non-sustainable pattern of use has prompted a series of corrective responses in the last 15 years. It is suggested that these should be taken further, with withdrawal of commercial wood supply from public lands. Under management regimes involving active participation by the local communities, the public lands may then be devoted to fulfilling the subsistence biomass needs of the rural population and commercial demands for non-wood forest produce.

MAN-MADE SAVANNAS

Over two-thirds of the uncultivated half of India today harbours tree and shrub savannas. These formations have all been derived from woodland ecosystems through a variety of human interventions (Singh et al. 1985; Gadgil and Meher-Homji 1985). In part this transformation has been deliberate, for savannas play an important support role in the nation's agro-ecosystems. Indian agriculture primarily involves the cultivation of cereal and legume crops coupled with maintenance of livestock. Only about a third of the fodder requirement of India's large livestock population is

221

met by crop residues, the other two-thirds must come from grazing on non-cultivated savanna lands adjacent to the cultivation. The consequent conversion of the vegetation on non-cultivated lands into dung is an important source of manure for the largely nutrient-poor tropical soils of India's farmlands. The livestock also supply much of the energy for agricultural operations and transport in the countryside. Furthermore, these savannas furnish fuelwood and dung cakes that fulfill the bulk of energy demands in rural India, along with grass for thatching huts, bamboos for basketry and house construction and a variety of other rural biomass needs (Ravindranath and Somashekar in press; Gadgil and Sinha 1985). Over centuries, therefore, India's rural inhabitants have created extensive savannas bordering their farmlands through extraction of woody biomass, grazing by livestock and annual dry season fires.

But this is not the only cause of the origin of Indian savannas. Vast tracts of woodlands often far away from cultivation have been converted into savannas through an overexploitation of woody biomass to meet urban industrial demands. These demands include timber and pulpwood, as well as substantial quantities of fuelwood. In fact, significant numbers of tribal peoples and rural poor derive a large part of their earnings from collecting and marketing fuelwood from forest lands at considerable distances from habitation.

India's savannas are now in a poor shape because the standing stock of their vegetation cover is being gradually depleted over vast tracts. There are of course exceptions. Some of the community-managed woodlots controlled by the village forest councils of Almora district in the central Himalayas or of Uttara Kannada district on the west coast are being managed in a sustainable fashion to meet village biomass needs (Somanathan 1991; Gadgil and Iyer 1989). Other degraded savanna lands are regaining their vegetation cover under good care by the village forest protection committees of West Bengal (Malhotra and Poffenberger 1989). Unfortunately, these are exceptions. Almost everywhere the cover is being depleted by the unregulated demands of both the commercial and subsistence sectors (Gadgil 1991).

CAUSES OF DEGRADATION

Table 11.1 attempts to trace the history of this pattern of overexploitation of India's vegetation cover and the consequent formation of large tracts of degraded savanna lands. At its root is the social system of resource use introduced by the British regime and subsequently strengthened after independence. This system aims to make biological and mineral raw materials available at exceedingly low rates for the commercial sector.

During the British regime this sector was primarily located in Great Britain, and woody biomass was devoted to the development of railways and shipping to facilitate the export of Indian raw materials along with direct export of timber, especially teakwood. Industry began to grow gradually in India after the First World War, developing rapidly after independence and thereby building up indigenous commercial demand for woody biomass. Three major devices have facilitated such cheap availability of biological raw materials:

(1) State takeover of all common lands as reserved and protected forest lands and revenue lands;

(2) Willingness of the State machinery to supply biomass produced on State-controlled lands at low prices, including a large element of subsidy introduced after independence; and

(3) maintenance of agricultural produce prices at a much lower level than industrial products.

Open access

This system of resource use has created a dual society on two different scales. Globally it has fostered a division between the industrial countries of the north and the developing countries of the south. At the same time, it has created within India a dual society made up of an elite, dominating the industry – organised services – intensive irrigated agriculture sectors, and the masses comprising landless agricultural labour, small farmers, rural artisans and tribals of the countryside and urban slum-dwellers making a living in the unorganised services sector. The masses have inadequate purchasing power and must perforce depend on biomass gathered free of cost from non-cultivated lands to meet a variety of their needs: fuel, thatch, small timber, fodder for their livestock and organic manure for their fields. Almost all these non-cultivated lands have been taken out of local-community control, a process that began with the British regime, and therefore serve as open-access common property resources subject to totally undisciplined harvests. These harvests serve not only the subsistence needs of the poor, but also provide them with income as does the marketing of fuelwood in towns.

Huge subsidies

Coupled to these non-sustainable harvests by the poor are the equally non-sustainable harvests by the forest-based industries. The imperative to

Table 11.1 The history of the overexploitation of India's vegetation cover

	Control exercised	Services rendered	Sustainability	Technical inputs
Pre-British period				
Irrigated agriculture	Land in theory belongs to Crown, controlled by local communities, cultivated individually, cannot be bought or sold	Surplus production as tax for the Crown, some production of commodities like cotton, subsistence consumption	Most likely to have been sustainable	Slowly evolving techniques of irrigation, mixed cropping etc.
Rain-fed agriculture	Land in theory belongs to Crown, controlled by local communities, cultivated individually, cannot be bought or sold	Some surplus production as tax for the Crown, largely subsistence consumption	Most likely to have been sustainable	Slowly evolving techniques of mixed cropping etc
Non-cultivated lands near habitation	Land in theory belongs to Crown, controlled and managed by local communities	Biomass inputs for agriculture, livestock and domestic sector	Most likely to have been sustainable, slow encroachment by agriculture	Promotion of economically valued species
Non-cultivated lands away from habitation	Land in theory belongs to Crown, largely unregulated usage except for specific resources such as elephants	Supply of commodities: timber for shipbuilding, spices like pepper, elephants for the army	Evidence of some non-sustainable usage as of exhaustion of elephant populations, slow encroachment by agriculture	None

(continued)

Table 11.1 *continued*

	Control exercised	Services rendered	Sustainability	Technical inputs
During the British regime				
Irrigated agriculture	Privately-owned land could be bought or sold	Tax in cash for the Crown, commodity production for the larger economy, some subsistence grain production	Problems like water-logging and salination	Enhanced irrigation based on larger dams
Rainfed agriculture	Privately-owned land could be bought or sold	Tax in cash for the Crown, some commodity production, subsistence grain production	Increasing problems of soil erosion	None
Non-cultivated lands near habitation	State-owned, mostly used by local communities under an open-access regime	Biomass inputs for agriculture, livestock and domestic sector	Exhaustive usage of natural biomass, encroachments for cultivation	None
Non-cultivated lands away from habitation	State-owned, but inadequately controlled, so that often subject to usage under open-access regime	Commercial timber and other commodities production	Exhaustive usage of natural stands, conversion to monoculture of species like teak and pine	So-called scientific forestry with little genuine technical content other than utilisation of species for railway sleepers or paper production

(continued)

225

Table 11.1 *continued*

	Control exercised	Services rendered	Sustainability	Technical inputs
Post-independence era				
Irrigated agriculture	Privately-owned land can be bought or sold	Commodity production for larger economy, consumption of commodities like fertilisers, pesticides, some subsistence grain	Problems of water-logging, salination, groundwater depletion and pollution, soil nutrient depletion, pesticide resistance	Green revolution technologies have substantial impact
Rainfed agriculture	Privately-owned land can be bought or sold; also illegally encroached government land	Production of subsistence grain and some commodities including pulpwood	Acute problems of soil erosion poorly developed	Dryland development technologies very poorly developed
Non-cultivated lands near habitation	State-owned, but used with inadequate control, subject to unregulated usage by local communities and by commercial interests	Biomass inputs for agriculture, livestock, domestic sector as well as for commercial sector	Rapid depletion of biomass, soil erosion, hardpan formation, encroachment for cultivation	Wastelands development technologies very poorly developed
Non-cultivated lands away from habitation	State-owned, but used with inadequate control by commercial interests and by local communities	Commercial timber, fuelwood and other commodities production	Exhaustive usage of natural stands, conversion to monocultures like teak, *Eucalyptus*	So-called scientific forestry with little genuine technical content other than utilisation of species in forest-based industry

promote rapid industrialisation following independence led to a system of extensive subsidies, which have been particularly extreme in the case of forest-based industry. Thus in 1958 the Karnataka State Forest Department permitted the West Coast Paper Mill to harvest bamboo at the rate of Rs. 1.50 per tonne, while the prevalent market rate was over Rs. 3000 per tonne. Today, in 1991, the Kerala State Forest Department is supplying *Eucalyptus* wood at Rs. 11 per tonne to the Hindustan Paper Corporation, when it costs the state at least Rs. 600 per tonne to produce this wood on State-owned land. This artificially cheap supply of raw materials has prompted growth of industrial demand well above the sustainable level. It has also encouraged the industry to go in for quick, short-term profits with little concern for the longer-term decimation of the resource base (Gadgil 1991).

The iron triangle

This system of subsidies encompasses a whole range of other resources as well (Figure 11.1 and 11.2). It has led to the elaboration of an "iron triangle", comprising the three components of the society involved in managing the subsidies. These are the politicians who decide on the subsidies, the bureaucrats who administer them, and the industry/organised services/ intensive agriculture sectors that benefit from the subsidies. The politicians and bureaucrats themselves largely belong to the industry/organised services/ intensive agriculture sectors. The result is that a narrow elite, about 15 per cent of India's population, corners most of the benefits of post-independence development efforts. This elite is shielded from the deleterious consequences, including the rapid degradation of vegetation cover in the non-cultivated lands, that have accompanied this non-sustainable pattern of development. Such costs are inflicted on the masses whose quality of life is intimately linked to the continued availability of biomass from the non-cultivated lands, the savannas covering a significant proportion of India's land surface.

CORRECTIVE RESPONSES

India is, however, a vibrant democracy and this iniquitous and non-sustainable pattern of development has stimulated corrective responses. These include a series of initiatives on the part of the State apparatus and of industry which have been triggered by protests by victims of environmental degradation. Predictably, the prescriptions by the different segments of Indian society differ widely, as summarised in Table 11.2. Those benefiting from the current patterns of resource use, the constituents of the iron

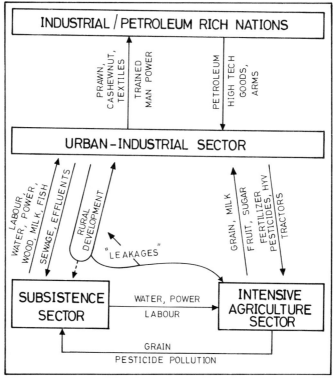

Figure 11.1 Internal patterns of resource flows and those linking India to industrial/petroleum-rich nations. Resource flows in India favour the more powerful interests and are responsible for exhaustive resource use in the countryside and the impoverishment of people dependent on resources in their immediate environment. The rural development programmes meant to correct these imbalances often fail to do so; instead, their benefits flow to those in power

triangle, naturally want minimal changes that primarily relate to an increasing role for private industry in the management of State-controlled lands (Shyam Sunder and Parmaeshwarappa 1987; Anonymous 1988).

Environmental activists committed to the need to bring about substantial changes in the present pattern of resource use belong to several different streams of thought. The dominant philosophy amongst them is Gandhian, with adherents such as Chipko leader Sunderlal Bahuguna (Guha 1989; Shiva 1991). This school romanticises pre-British agrarian society and is opposed to the process of industrialisation on the Western pattern. The other significant group of environmentalists might be termed appropriate technologists. This school, represented by scientists like Amulya Reddy, favours industrialisation along a more equitable and sustainable path (Goldemberg et al. 1988).

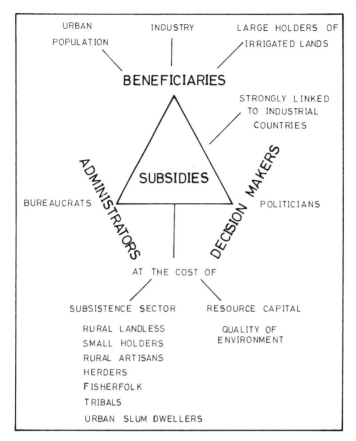

Figure 11.2 The iron triangle governing resource use patterns in India. Large State-sponsored subsidies have created an iron triangle of components of Indian society benefiting from, administering and deciding upon resource subsidies. Constituents of this iron triangle are forcing the country into a pattern of exhaustive resource use at the expense of the environment and the majority of the people

From this debate a number of positive measures have emerged. These include:

(1) Beginning with the Fifth Five Year Plan of 1975–1980, it is compulsory for all major development projects to carefully prepare an environmental impact statement and to provide for compensatory afforestation in the event of loss of any forest cover.

(2) The *Forest Conservation Act 1980* prescribes that no land should be alienated from control by the forest department without the consent of the central government. This has largely stopped all such alienation

Table 11.2 Current resource-use trends and prescriptions by the different components of Indian society

	Current trends	Prescriptions by iron triangle constituents	Prescriptions by Gandhian environmentalists	Prescriptions by advocates of appropriate technology
Extent of resource use	Moderate and growing by constituents of iron triangle. Low, often declining by others	Promote intensified resource fluxes channelled towards constituents of the iron triangle	Aim at very moderate resource use, by all people; halt further intensification of resource use	Encourage increasing resource use while reducing disparities and ensuring sustainability
Control over resources	Increasing control by State apparatus	Increasing control by State apparatus and business corporations	Highly decentralised control by local communities	Decentralised control by local communities, with strong input from technical experts, NGOs*
Sustainability	Sequential resource exhaustion at the cost of those with little control over resources	Little serious concern over issue of exhaustive usage	Must be ensured by halting resource use intensification, coupled with local community control	Must be tackled by promoting disciplined resource use, coupled with technical inputs
Role of science and technology	To extend reach over ever-larger range of resource types and spatial domain for those in control of resources. To rationalise this process to the whole society	Continue technical inputs to promote current trends	Science and technology viewed as a negative influence	Need to reorient science/technology inputs to inject sustainability and participation by wider segments of society

*NGO = non-government organization

and official statistics indicate an essential halt to the process of deforestation. While, unfortunately, this is not entirely true, the Act has definitely led to a deceleration of forest cover loss.

(3) Beginning in the late 1970s the absurdly high levels of subsidies enjoyed by forest-based industry have been gradually reduced, thereby helping to motivate the industry to use resources more carefully. Thus, in Karnataka, the supply rate of bamboo for paper mills has been increased from Rs. 1.50 per tonne to Rs. 600 per tonne, although this is still well below the market price.

(4) At the same time, the farmers have been encouraged to take to tree-crop production as a part of the social forestry programme, and many industries and farmers have developed healthy links for fulfilling the needs of forest-based industries.

(5) Since the 1970s and beginning with the Village Forest Protection Committees of West Bengal, local tribals and villagers have been given an economic stake in the well-being of local forest resources. The response has been most encouraging. A major step in extending such systems to other States was taken by the Ministry of Environment and Forests, Government of India, in June 1990. Many State governments including West Bengal, Haryana and Gujarath were motivated to initiate measures to create village community-based management systems for involving the local people positively in the sustainable use and restoration of local forest resources.

(6) Beginning in the late 1970s with the State of West Bengal and the mid-1980s with Karnataka, many State governments have been decentralising political decision-making. The village and district level political authorities thus created are far more motivated to ensure prudent use of local resources. Other States are expected to follow suit in the decentralisation of power, creating more favourable conditions for the better management of forest resources.

(7) The National Wastelands Development Board created in 1985 has taken novel initiatives in decentralising integrated planning of the use of natural resources. It is spearheading efforts at restoring the vegetation cover of vast stretches of degraded savanna lands by meaningfully involving the local people.

(8) The amended National Forest Policy 1988 is a major step in recognising the primacy of the need to fulfill the subsistence biomass requirements of local people from the non-cultivated lands.

AN ALTERNATIVE REGIME

What is now needed is to take this process further and implement it more vigorously at the ground level. A comprehensive package of policies to accomplish this is developed in Table 11.3. They have the following three aims. Firstly, the phased transfer of all wood production for commercial purposes to privately-owned lands. India has 150 million ha of land under cultivation, but the bulk of food production comes from some 30 million ha of intensive agriculture, primarily in Punjab, Haryana, Western Uttar Pradesh and Andhra Pradesh. A good proportion of the 150 million ha of cultivated lands is hilly terrain or land in tracts of low, uncertain rainfall that would be far better put under tree cover. Tree production has been shown to be far more efficient on private land than on government-controlled forest lands. Thus, in Karnataka, the annual productivity of government *Eucalyptus* plantations is as low as 1.5–2 tonnes per ha, while that on private land is 10–15 tonnes per ha. At these levels only about 3–6 million ha, less than 4 per cent of all cultivated land, would be adequate to meet India's total commercial timber and pulpwood needs of less than 30 million tonnes per year. By eliminating the need for any wood harvesting from the 60 million ha of reserve forest lands, whose current average wood productivity is just 0.5 tonnes per ha, this switch would permit a very substantial build-up of tree cover on these lands. Even while generating wood for commercial purposes on private lands, it would be possible to use promising species like bamboos which can yield large, sustained quantities of excellent raw material for the paper and polyfibre industries without clear-cutting the stands.

Secondly, the subsistence biomass needs of tribal and rural populations should be met from a network of community lands close to their settlements. Such lands should ideally be managed by relatively small and homogeneous communities such as village hamlets with 50–100 households. These user groups need to be properly empowered and helped to establish regimes of replenishment and sustainable use over small, specifically assigned areas drawn from revenue wastelands, protected forests and reserve forests. A vegetative cover of a diversity of species could be developed and harvested in the traditional mode that involves use of leaf and branch loppings, fallen and dead wood, grass and fruit collection, but no tree felling. These community woodlots could include small patches of totally protected vegetation on the model of the traditional system of sacred groves. A very substantial and diverse tree cover could be built up on some 10–20 million ha of savanna lands close to habitation under such a management regime.

Thirdly, the remaining 50–60 million ha of reserve forest lands should be brought under a system of production restricted to non-wood forest produce of commercial value. The diverse tropical forest cover of India is

232

Table 11.3 Proposed pattern of management of biomass resources from India's non-cultivated lands

	Non-cultivated lands adjoining human settlements	*Non-cultivated lands away from human settlements*
Current ownership status	State-owned, often with user privileges but no rights of regulation assigned to local people; occasionally owned privately	State-owned with few or no user privileges and absolute rights assigned to local people
Current dominant usage pattern	Unregulated biomass harvests by local people and their livestock; encroachment for cultivation leading to degradation of land, vegetation and biodiversity	Non-sustainable harvests to fulfill urban-industrial demands at highly subsidised rates, not only from outside but sometimes from within nature reserves all out attempts to exclude local people
Proposed management pattern	Well-organised community-based management with land ownership vested in the State special incentives for conserving standing biomass	State-managed with active involvement of local communities in sustainable use of non-wood forest produce and in conservation of biodiversity
Expected vegetation cover	Good standing crops of indigenous multiple-use trees, shrubs and grasses with nuclei of protected forests	Good standing crop of indigenous tree species enriched with a variety of species yielding non-wood forest produce and larger strictly protected and nature reserves
Expected economic services	Fulfilment of biomass needs of tribal/rural population; surplus, if any, for urban-industrial sector	Non-wood forest produce for industrial needs; support to tribal/rural economy
Potential for employment generation	Considerable labour inputs would be called for in the development, proper maintenance and utilisation of community woodlots	Substantial additional employment in harvesting, processing and marketing of non-wood forest produce
Expected benefits for biodiversity conservation	Protected species, diverse nuclei of community woodlots, sacred trees and small sacred groves	Larger nature reserves well protected via local co-operation and removal of commercial pressures; much more diverse composition of forests generating non-wood forest produce
Other environmental services	Maintenance of tree cover on non-cultivated lands near human settlements; better watershed protection; higher levels of carbon sequestration	Maintenance of natural vegetation with good cover on all land away from human settlements; better watershed protection; higher levels of carbon sequestration

exceedingly rich in species yielding such produce; in some cases, as with the leaves of *Diospyros* used for making country cigarettes or beedis the revenue accruing to the State from the collection of these commodities far exceeds timber revenue from the same localities. At present these non-wood forest products are managed poorly; rights are auctioned off to contractors who have no stake in long-term sustenance of the produce and who pay local tribals and villagers extremely low rates for collecting it. It is necessary to systematically encourage and enrich the bulk of India's forest tracts by species that yield such non-wood forest produce. It is also necessary to create management systems that will generate a genuine economic stake for local inhabitants in ensuring that forest stocks do get replenished by such species and that they harvest, process and market the produce through institutions that bring to them the full value of that produce.

Such a management regime is certainly feasible in that the country's savanna lands, as well as natural and man-made woodlands, are quite capable of attaining the requisite levels of biological productivity. At these levels, biological production would be adequate to meet, not just the current biomass demands of the commercial and subsistence sectors, but also the projected increases in these demands over the next 20 years. The proposed regime would additionally serve to greatly enhance the ecological services that the non-cultivated lands provide, and create sorely needed employment opportunities and incomes for rural and tribal communities. Equally importantly, it would confer on them dignity by seeking their active participation in the management of natural resources.

This does not mean that all elements of the proposed alternative regime would be readily accepted; for there are strong vested interests favouring the perpetuation of the current patterns of resource use. These include the interests of the bureaucracy against giving away any power over resources currently in their hands, as well as the interests of the elite in continuing to enjoy access to natural resources at high levels of state subsidies. However, there are hopeful signs that forces favouring more sustainable and equitable patterns of resource use are gathering strength. It is likely then that several elements of the package of policies outlined above will be implemented in years to come, restoring the productivity of India's savanna lands.

ACKNOWLEDGEMENT

Department of Environment, Government of India, has generously supported this work over the last 9 years.

REFERENCES

Anonymous (1988) *Whither common lands* ? Samaj Parivartana Samudaya, Dharwad.

Gadgil, M. (1991) Restoring India's forest wealth. *Nature and Resources* 27 (2):12–20.

Gadgil, M. and Iyer, P. (1989) On the diversification of common property resource use by the Indian society. In: F. Berkes (ed) *Common property resources: Ecology and community based sustainable development.* Belhaven Press, London, pp. 240–55.

Gadgil, M. and Meher-Homji, V.M. (1985) Land use and productive potential of Indian savannas. In: J.C. Tothill and J.J. Mott (eds) *Ecology and management of the world's savannas.* Australian Academy of Science, Canberra, pp. 107–13.

Gadgil, M. and Sinha, M. (1985) The biomass budget of Karnataka. In: C.J. Saldanha (ed) *Karnataka State of Environment Report 1984–85.* Centre for Taxonomic Studies, Bangalore, pp. 19–30.

Goldemberg, J., Johansson, T.B., Reddy, A.K.N. and Williams, R.H. (1988) *Energy for a sustainable world.* Wiley Eastern Limited, New Delhi.

Guha, R. (1989) *The unquiet woods: Ecological change and peasant resistance in the Himalayas.* Oxford University Press, Delhi.

Malhotra, K.C. and Poffenberger, M. (ed) (1989) Forest regeneration through community protection: The West Bengal experience. Proceedings of the Working Group Meeting on Forest Protection Committees, June 21–22. West Bengal Forest Department, Calcutta.

Ravindranath, N.H. and Somashekar, H.I. (1992) Biomass production, utilisation and conservation in a semiarid area. In: C.J. Saldanha (ed) *Karnataka State of Environment Report 1991–92.* Centre for Taxonomic Studies, Bangalore.

Shiva, V. (1991) *Ecology and the politics of survival.* Sage Publications, New Delhi.

Shyam Sunder, S. and Paramaeswarappa, S. (1987) Forestry in India – The forester's view. *Ambio,* 16 (6):332–7.

Singh, J.S., Hanski, Y. and Sajise, P.E. (1985) Structural and functional aspects of Indian and Southeast Asian savanna ecosystems. In: J.C. Tothill and J.J. Mott (eds) *Ecology and management of the world's savannas.* Australian Academy of Science, Canberra, pp. 34–51.

Somanathan, E. (1991) Deforestation, property rights and incentives in Central Himalaya. *Economic and Political Weekly,* 26 (4):37–46.

Section 5

Private-property grazing systems

CHAPTER 12

LAND USE IN VENEZUELA

J.F. Silva and A. Moreno

SUMMARY

The availability of soil moisture during the dry season and the low-level nutrient status of the soil are the two main ecological constraints limiting natural and agricultural productivity within the Orinoco savannas of Venezuela.

Contrary to the perception expressed in some studies, the agricultural frontier is not expanding in the Venezuelan savannas. Instead, agriculture is still largely underdeveloped and marginal, and the savannas remain essentially uncolonised. The main land use is extensive cattle-raising under low-input management. Transformation into land suitable for crop production can only be achieved with considerable investment in infrastructure, roads, machinery and large subsidies. Crop production is technically possible but in most areas not consistent with national economic interest. The general low productivity of seasonal savanna systems needs to be recognised. In the last 30 years national agriculture, despite many government subsidies, programmes and substantial investment, has not been able to compete with imports.

A strategy of use based on extensive cattle-raising, coupled to technically based improvement of the quality of native pastures and the remodelling of the trading system, would allow productivity to improve without major environmental risks. Tourism and recreation, as well as the development of ecologically sound industries to exploit native animal populations, could also help to increase productivity and would be self-sustainable.

INTRODUCTION

Venezuela has nearly 300,000 sq km of savanna lands, which represent almost one third of the national territory. Most of this is located in the

239

Orinoco Llanos region in the central part of the country. Although there is an important amount of ecological information on the Venezuelan savannas (Sarmiento 1983, 1984; Acevedo and Silva 1985), studies on savanna land use are restricted to some areas. There is a lack of research linking the ecological and economical aspects of land use with a more general approach.

In this chapter, we present a preliminary view of savanna land use in Venezuela, and try to emphasise similarities and differences with the rest of the country. The influence of ecological constraints and government policies in the last three decades is then examined. The analysis is restricted by the lack of specific information on savanna lands, since official statistics refer to administrative units (States), not savanna land units. The statistics are scarce and have been further reduced during the last decade. Most of the information presented is based on the Anuario Estadístico de Venezuela and on the National Censi of 1961 and 1971.

Since this is a preliminary view, we concentrate on two representative areas within the Orinoco savannas: (1) the Mesas Region, located in the States of Anzoategui and Monagas in the eastern part of the country, and (2) the Llanos of Apure Region, which comprises most of the State of Apure in the south of Venezuela (see Figure 12.1). We selected these two areas because they present extended and continuous savanna vegetation, comprising a gradient of ecological conditions, from drier and well-drained seasonal savannas to wetter, periodically drenched hyper-seasonal savannas. Furthermore, a very important oil extractive industry is located in the Eastern Mesas Region.

These two areas cover nearly 80,000 sq km, about one-third of the total savanna area in the Orinoco Llanos and almost one-tenth of the national territory. In 1982, the total population of the three States, Anzoategui, Monagas and Apure, was 1,366,499 people. This is equivalent to 8.7 per cent of the total national population; 80 per cent of the people live in cities. The main cities of the Eastern Mesas are Maturin, El Tigre, El Tigrito and Barrancas. In Apure, the main cities are San Fernando, Achaguas, Mantecal and Elorza.

THE ORINOCO SAVANNAS

Geomorphology

The Orinoco Llanos are located in a large geosyncline, limited by the Guiana Shield to the south, the Andean Cordillera to the west and the

240

Figure 12.1 The Orinoco Llanos Region of Venezuela (dashes) and the location of the two study areas, Apure and the eastern mesas of Monagas and Anzoategui (dotted)

Caribbean Cordillera to the north, and comprise around 500,000 sq km of quaternary sediments. Pleistocene tectonics raised some areas, isolating them from further depositions and exposing them to erosive processes (Mesas Region). Other areas were depressed and progressively covered by more recent sediments.

Climate and soils

In contrast to the diversity of landscapes due to a variable geomorphology, the Orinoco Llanos have a homogeneous, macrothermic and isothermic climate. Annual mean temperatures are 24–28°C. Biotic activities are never limited by low temperatures. The region is wet, with annual rainfall varying from 900 mm in the Mesa de Guanipa in the Eastern Mesas to 1800 mm in the Andean piedmont. Rain is very seasonal, with a wet season from May to November, followed by a dry period with no rains. All savanna soils can be considered as homogeneously distrophic as a result of their origin and the intense leaching (Sarmiento 1983). The availability of soil moisture during the dry season and the low-level nutrient status of the soil are the two main ecological constraints that limit natural and agricultural productivity in this region.

The accumulation of dry biomass in the grass layer makes possible the annual occurrence of fires during the dry season. Fire removes the standing biomass, promotes the growth of native grasses (Silva et al. 1990, 1991), and reduces tree growth (San José and Fariñas 1983; Fariñas and San José 1985). Most of these fires are man-induced.

In the Mesas Region, seasonal savannas are located on the raised flat lands, locally known as mesas, from the lower Pleistocene age. These have been fragmented and dissected, and present a flat or gently rolling topography. The annual rainfall, between 900 and 1000 mm, is the lowest of any savanna in the Llanos Region. Soil erosion is important in both hydric and eolic forms. Soils are oxisols and ultisols, very poor in nutrients, with good drainage and a deep water-table (COPLANARH 1974). They are never flooded and after the rains they dry rapidly. Standing crop figures for these savannas are in the range of 300–700 g/m^2 (Sarmiento 1984).

The savannas from the State of Apure are on alluvial overflow plains and on eolian plains (Sarmiento et al. 1971). The former date from late Pleistocene to early holocene. The overflow of rivers has modelled the landscape, giving rise to a network of physiographic units such as pediments, deposition and drainage basins, with fine sediments (silt and clays). Most soils are alfisols or inceptisols, and vertisols are found at the bottom of the topographic catenas (Comerma and Luque 1971; MARNR 1982). The eolian plains are considered to be remnants of a former arid morphogenesis from the Würm glaciation. They are characterised by extensive dune fields superimposed on larger areas of loess-like material, partly covered by younger alluvia. Soils are very sandy (psamments), but topography may create restrictions to drainage.

In the Apure savannas, soils are also poor in nutrients, but alluvial soils are nutritionally better than those from the mesas. Standing crops in those savannas ranges from 700–800 g/m^2 (Sarmiento 1984).

Physiognomy is very similar in both regions: open savannas and grasslands are predominant. In hyperseasonal savannas, trees are restricted to better-drained river banks or dunes. In the better-drained, seasonal savannas, trees are restricted to areas where the water-table is reachable.

Flora and fauna

Both regions have similar flora. Common tree species are: *Curatella americana, Bowdichia virgilioides, Byrsonima crassifolia* and *Casearia syl-*

vestris. The grass layer is dominated by species from the genera *Trachypogon, Andropogon, Axonopus, Paspalum, Leptocoryphium, Sporobolus,* and *Elyonurus,* and by sedges from the genera *Bulbostylis, Cyperus, Dichromena, Eleocharis, Fymbristilis, Scleria* and *Rynchospora.*

Within the Orinoco savannas there are two large national parks. One (Aguaro-Guariquito), was created in 1974 and is located in the State of Guarico, out of the study area. It is the second largest in the country and covers 560,000 ha (MARNR 1978). The other, Santos Luzardo, was created in 1988 and is located in the State of Apure. It covers 384,368 ha. Both parks have a very diverse wild fauna, especially birds. They lack any infrastructure for tourism or educational purposes, and even road access is difficult during the rainy season. The potential for fauna-based ecotourism is great, however.

Although low population density, remoteness and lack of access protect the biota and the ecosystems in these savannas, many vertebrate species have experienced heavy pressure from hunting and poaching. Of all the vertebrates, 94 species are reported as needing special protection. Of these, 11 are listed as highly endangered. The most well known of these include the cats (*Panthera onca* and *Felis pardalis*), the jabirú (*Jabiru mycteria*), the scarlet and the red-bellied macaws (*Ara macao* and *A. manilata*) and the Orinoco's caiman (*Crocodylus intermedius*) (MARNR-BIOMA 1991). These, and other animal species, are already under protection.

People

Two main aboriginal groups are found in the area of study (Mérida 1966). The first group are found in the Mesas of Anzoategui and are known as the *Cariñas.* This small group of people are almost totally integrated to Spanish culture and live as peasants. The second group live in the Apure and are known as the *Yaruros.* Despite the intense pressure from Creole settlers, the *Yaruros* still maintain some traditional aboriginal land-use practices. Both groups are much endangered, not only because of cultural pressures, but also because of disease and direct attack from Creole settlers.

SAVANNA LAND AGRICULTURE

Savanna land use

In the three States under study, a high proportion of the land is covered by native grasslands and is used mainly for extensive cattle-raising with traditional, very simple management techniques. Only a small fraction of

Table 12.1 Land use as a percentage of total land devoted to crops or introduced pastures, savanna rangelands and other (including fallow lands, forests, swamps, etc.)

	State		
	Monagas	*Anzoategui*	*Apure*
Agriculture			
Crops	10	7	2
Improved pastures	16	16	4
Savanna	47	38	72
Other	27	39	22

Source: National Census of 1971

the land is used for agriculture, including cultivated pastures. Agriculture is least developed in the State of Apure (see Table 12.1).

Until very recently, most of the ranches did not have any fencing. Extensive cattle-raising is conducted on native, low-nutrition grasslands, with little technical input. Maximum carrying capacity is only 0.5 animal units per ha. Savannas are seasonally grazed, and in hyperseasonal savannas the herd is moved toward better-drained areas during the peak of the rainy season. At the beginning of the dry season, cattle are sold to intermediaries who move them to ranches in the piedmont. There, the animals are fed on cultivated pastures for a few months before sale to meat industries.

Use of modern agricultural technology is very limited. For crops, like cotton, there is a very moderate technical support (irrigation, fertilisers, pesticides). Farming has low-profit margins, partly because of government policies controlling the market prices of agricultural produce and partly because most of the income generated goes to the intermediaries and only in a small degree to the producers (CENDES 1975). Furthermore, an important fraction of the profits, rather than being used to improve savanna productivity is diverted to other activities. As a result, these ranches exhibit a general situation of low productivity and progressive decapitalisation (Martinez 1975).

Savanna land tenure

The pattern of land use described above conforms with a system of ownership characterised by very extensive ranches (see Table 12.2), most

Table 12.2 Distribution of land by size of holding

Size	Apure		Monagas		Anzoategui	
ha	n	ha	n	ha	n	ha
<1	649	276	306	150	651	324
1–19.9	7490	37,311	11,659	65,523	10,937	47,203
20–99.9	1729	54,939	1841	68,835	1885	74,695
100–499.9	383	79,390	57	119,020	1144	270,330
500–2499.9	290	321,058	243	239,835	781	784,495
>2499.9	412	3,729,305	74	388,131	151	758,411
Total	11,062	4,729,305	14,774	881,494	15,843	1,949,458

Source: National Census of 1971

Table 12.3 Distribution of land tenure by area. Source: National Census of 1971

	Anzoategui	Monagas	Apure
Owned (%)	85	83	83
Rented (%)	2	3	1
Occupied (%)	12	12	15
Other (%)	1	1	1
Total area (ha)	1,949,458	881,494	4,222,309

of which (80 per cent) are privately owned (see Table 12.3). Both the tenure system and the type of use have been predominant since colonial times (Rodriguez Mirabal 1987) and, despite the agrarian reform of 1961, have not been changed.

As shown in Table 12.2, in the State of Monagas only 2 per cent of the farms are larger than 500 ha, but they represent 71 per cent of the total area. In Anzoategui and Apure, 6 per cent of the farms are larger than 500 ha, but they represent 80 per cent of the total area in the former and 96 per cent in the latter. Note that these statistics are for the whole of each State; the concentration is even higher in the savanna lands since small farms are more abundant in the forested mountains. Size and type of land use are closely related. A high proportion of farms below 20 ha used exclusively for crop production, whereas almost all farms larger than 500 ha are devoted entirely to the production of cattle.

Concentration of land property has been considered one of the major obstacles to agricultural development within the country. Since large farms are mostly devoted to extensive cattle-raising, crop producers are restricted to small farms. Some people consider this decreases the efficiency of crop production as large landowners are not prepared to subdivide their land and sell part of it to farmers. As a result they are forced to rent the land they use in a way that increases management costs and risks and also makes it more difficult to obtain credit. The incentive to improve or maintain the land is less under such arrangements (Orta 1974b).

The agrarian reform

The land tenure system, characterised by very large landed estates (*latifundios*), is considered one of the great obstacles to the formation of agricultural capital and to technological innovation (Orta 1974b). The Agrarian reform was approved as a law in 1961, as a demonstration of the political decision of the ruling democracy to transform the rural system. The goals were to bring about a more equitable distribution of land ownership by breaking up large estates and giving the rural population a significant role in social and economic development and in political affairs. It was anticipated that this initiative would provide 300,000 rural families with 10 ha of land in property and raise the living standard of rural workers. The technical and financial assistance to facilitate this was also to be provided.

In implementation, however, the programme was very different and, after the first decade of enforcement, the programme was progressively abandoned. The reasons for failure were complex but relate largely to difficulty in breaking up large estates. Instead of doing this, most of the land allocated was public land with little production potential. Less than 10 per cent of all the land provided to the peasants came from privately-owned land. Thus, the major estates remained untouched. Furthermore, only a third of the proposed number of families received land, and of those 50 per cent received lots of 5 ha or less. Only 20 per cent received financial support.

The lack of technical and financial aid, and the limited area of agricultural land allocated to each peasant, made it very difficult to accumulate and consolidate working capital. A chronic debt problem resulted and to clear these debts families were forced to sell their land and migrate to the cities in search of jobs. Ironically, this land was then purchased by large estate owners with the result that land ownership became more concentrated and more public land entered into private ownership. The outcome of agrarian reform was the exact opposite of what was intended.

Traditional crops

The main crop in the Orinoco savannas is corn. Other traditional crops are beans, manioc and cotton. A significant fraction of agricultural production is for the national markets and the agro-industry, only a small

Table 12.4 Percentage of agricultural produce consumed on the farm. Source: National Census of 1971

Proportion of produce consumed on the farm	State		
	Anzoategui (%)	*Monagas* (%)	*Apure* (%)
All	5	1	2
Major portion	22	7	19
Minor portion	54	71	67
None	19	20	12

fraction being consumed on the farms (see Table 12.4).

During the past 3 decades, traditional crops have shown important fluctuations, both in harvested surface and production. Despite increasing needs traditional crop production either stagnated or declined. This picture contrasts with an apparent growth of agricultural production in the country, which parallels population growth from 1968 to 1980. Most of the production increase has been in rice, sorghum, eggs, poultry, fruits, pork, meat, milk and vegetables. These items are cultivated in the central and western parts of Venezuela, on lands from dry and wet forests, and in fertile valleys from the mountains and the Llanos regions.

One remarkable case is cotton, which almost doubled in harvested land during the early 1970s to decrease to very low levels subsequently (see Figure 12.2). From being a very important crop in the State of Anzoategui, it declined considerably during the 1970s to reach the low levels of Monagas and Apure.

Bread made of corn is an important element in the diet of Venezuelan people. Corn is also used to produce cooking oil. Corn production, however, has shown very wide fluctuations, both nationally and within the Orinoco savannas (see Figure 12.3). In contrast, corn imports have increased from very low levels in 1969 to a level that is two times higher than national production (see Figure 12.4). Competition from imported wheat has been an additional problem for corn producers. In 1951, 65 per cent of cereal consumption was provided by corn and 28 per cent by imported wheat. In 1984, the share of corn had declined to 37 per cent of the total, and that of wheat had increased to 50 per cent (Hernandez and Merz 1986).

Therefore, despite direct and indirect subsidies to agriculture, traditional crops have shown stagnation or deterioration during the last 30 years. There is no apparent direct relationship between Venezuelan agricultural production and market prices, and other factors, such as certificated seed supply, storage capacity, marketing assurance, technical assistance and competition with internationally produced substitutes like wheat, may be playing an important role (Kastner et al. 1983).

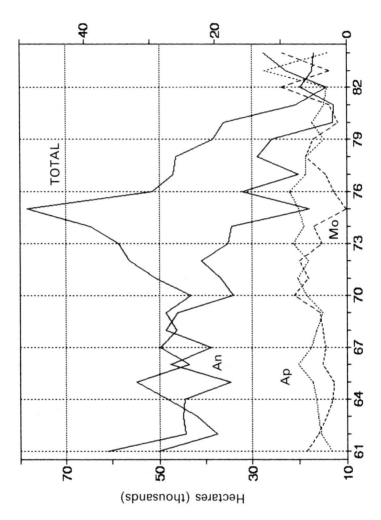

Figure 12.2 Harvested surface for cotton during the period 1961–1983, for the country (total) on the left vertical axis, and for each of the three States, Anzoategui (An), Monagas (Mo) and Apure (Ap), on the right vertical axis

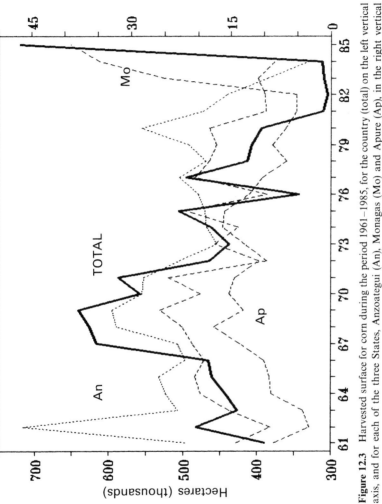

Figure 12.3 Harvested surface for corn during the period 1961–1985, for the country (total) on the left vertical axis, and for each of the three States, Anzoategui (An), Monagas (Mo) and Apure (Ap), in the right vertical axis

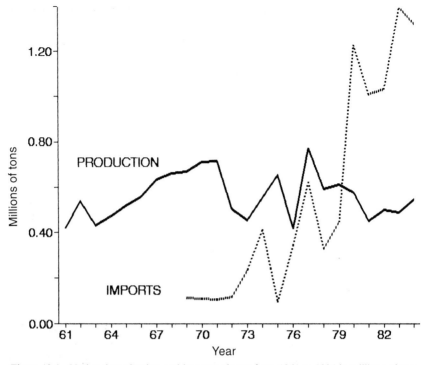

Figure 12.4 National production and imports of corn from 1961 to 1984 in millions of tons

RECENT AGRICULTURE IN SAVANNA LANDS

In addition to the support for traditional crop production, government policies have sought to introduce new crops that are adapted to local conditions, as well as other economic programmes to enhance the productivity of savanna lands.

The peanut plan

One example of this approach has been the peanut programme targeted to Mesa Guanipa within the savannas of the State of Anzoategui. The introduction of this crop was facilitated by a concessional loan programme designed to help retired employees from the oil companies, store owners and young professionals from the main towns (El Tigre, El Tigrito) to become farmers. The programme included cooperation with agro-industry and technical assistance to improve production as well as trading. A cooking oil factory, and several organisations for technical assistance and trade, were established (CENDES 1975).

Peanut production under this programme makes intensive use of irrigation, fertilisers, pesticides and machinery in a way that permits farmers to grow two crops per year. Nevertheless, the cooking oil industry finds it cheaper to import peanut oil. As a result, producers have only been able to persist because they continually receive government subsidies.

Sorghum

Sorghum was introduced in the savannas of the Mesas Region as well as in the deciduous forests of the Central Llanos Region during the early 1970s. The harvested surface increased considerably during the 1970 decade, both nationally and in the Mesas, but for 1984 the latter had declined to less than 20 per cent of the 1980's peak.

Pine plantations

Another programme was the pine plantations in the southern mesas of the State of Monagas, sponsored by the Central Venezuelan Government's Corporation of the Guayana Region. It was initiated in 1968, with the goal of sowing 140,000 ha with Caribbean pines at a rate of 10,000 ha per year for the paper industry. By 1985, the programme was to provide two-thirds of national paper needs, by means of a paper factory to be installed in the area, in partnership with an international consortium.

The plantations have developed well, reaching nearly 70 per cent of their goals. However, the paper mill has not been installed and it does not seem to have any priority within the current plans of the Central Venezuelan Government. Meanwhile, the pines are used for lumber and resins, in small-scale operations. The huge investment provided for the plantations is now at risk due to the delay and expected cancellation of the plans to build the factory.

Flood plain management

Another government programme was started in the early 1970s to improve the carrying capacity of the flood plains savannas in the State of Apure. A system of dikes, to control the flooding and to keep a reservoir of water for the dry season, was constructed on an experimental basis. Several research programmes followed the changes in floristic composition, nutrient dynamics and ecological status of the lands affected, as well as animal production (Gil Beroes 1976; Lopez-Hernandez et al. 1983; Berrade and Tejos 1984; Tejos 1984). The results showed a significant increase in the contribution of several nutritious species (*Leersia hexandra, Hymenachne amplexicaulis,* etc), higher animal production and changes in the carrying capacity from 6 to 1.5 ha per animal unit. However, the programme has been practically abandoned as a result of political changes and, apart from the results mentioned above, no evaluation has been made of the social and economic implications of the programme.

MARGINALITY OF SAVANNA AGRICULTURE

Savanna lands in Venezuela, like those found in most other parts of the world, are of limited production potential. Under Spanish rule, the economy was orientated to crop production for international markets, and the country was not structured as a geographical and social entity. Instead, each independent region was connected to international trade by a city port. Whilst the Andean region experienced important economic prosperity, first from wheat agriculture and later from coffee plantations, the growth of the Barinas region was promoted by tobacco exports and the Central region prospered due to the cocoa plantations in the Coastal Cordillera, the Orinoco Llanos, and especially the savannas, remained socially and economically depressed. This stage lasted until the crisis of the agro-export economy, during the worldwide depression of the 1930s.

Both the nutritional conditions of the soils and the strong seasonality in water availability are important natural constraints on the development of sustained agriculture in these savannas. Crop productivity is consequently low and any increase can only be achieved with large subsidies and considerable investment in infrastructure of roads, machinery and capital improvements.

Savanna lands have, therefore, remained marginal to the agricultural development of the country, which is taking place instead in those regions with higher natural productive potential, such as the western piedmont of the Llanos Region with dry and mixed forests presently being replaced by annual crops. This marginal condition of savanna lands is not new. It originated in colonial times, and there is evidence that it is now increasing as a consequence of the advance of the agricultural frontier in other regions.

During the last 2 centuries (agro-exporting period), these savannas contained most of the national herd and were the major providers of meat and milk, based on extensive, semi-pastoral techniques. During the last 30 years, the agricultural frontier has advanced and replaced very extensive forests in the piedmont and the southern Maracaibo (Zulia) regions. These

Table 12.5 Changes in the size of the cattle herds in the three States of the study area, in the whole country, and in the State of Zulia for comparison (× 1000)

	1937	1950	1961	1971	1984
Apure	954.7	1154.3	1315.7	1283.6	1782.4
Anzoategui	352.7	432.1	387.1	436.1	524.6
Monagas	136.0	278.9	196.2	302.8	363.6
Zulia	289.1	502.4	908.8	1792.8	2645.5
Venezuela	4303.5	5768.8	6440.7	8549.3	11,843.6

Source: Anuario Estadístico Agropecuario

replacement systems have become stable as grasslands with introduced pastures, connected to livestock industrial development. As a result, national needs for these foods are provided in a substantial proportion by these new agricultural regions. As an example of these tendencies, the contribution of the State of Zulia to the national cattle herd has increased very rapidly, whereas those of the states of Apure, Anzoategui and Monagas have only increased moderately since 1935 (see Table 12.5).

The existence of numerous oil wells in the Mesas Region has not had any positive impact upon local economies nor upon agricultural development. On the contrary, this has further contributed to the abandonment of rural areas.

In summary, and contrary to the perception of some studies (Cochrane et al. 1985), the agricultural frontier is not expanding in the Venezuelan savannas. Instead, agriculture is still largely underdeveloped and marginal in both poorly drained and well drained savannas of the Orinoco Llanos, and they remain essentially uncolonised.

THE FRAMEWORK OF VENEZUELAN AGRICULTURE

The lack of precise information, the complex web of policies and programmes, and the lack of coincidence between the proposed goals and the achievements make analysis of recent Venezuelan agricultural history very difficult. On the other hand, this paper is not intended to clarify the causes of the present conditions of Venezuelan agriculture. Instead, we will highlight a few main aspects of its development.

During the last three decades, around 20 per cent of oil revenue has been directed towards the agricultural sector, as part of a strategy to reduce food imports by promoting the development of a food industry based on national agricultural production. This strategy has required the development of a large and complex set of institutions to channel public and private investments, including banks (Banco Agrícola y Pecuario, Banco de Desarrollo Agrícola), national corporations (Corporación de Mercadeo Agrícola), regional corporations (Corpo Andes, CorpoLlanos, etc). Furthermore, major investments have been made to develop the infrastructure needed for modern agriculture, such as irrigation systems, electrification of rural areas, roads and telephone systems, etc.

Besides the inefficiency of investment in agriculture, especially major projects like irrigation systems, the agricultural sector has not been able to consolidate its assets and has tended to divert profits towards other sectors, instead of reinvesting in its own development (Orta 1974a).

The policy of subsidies

The six previous democratic administrations have maintained a constant policy of direct and indirect subsidies for agriculture. Among the former, the prices of fertilisers and pesticides have been maintained much lower than in the international market, via direct government subsidy of the petrochemical industry. This has induced the use and abuse of these chemicals, creating serious pollution problems in some regions with intensive crop agriculture.

Among the latter, the prolonged overvaluation of the national currency (bolivar), coupled with significant quantities of foreign currency from oil exports, has made possible the importation of agricultural machinery and other insumes at reduced prices. This overvaluation, however, has made Venezuelan products much less competitive in the international markets, and reduced agricultural exports to minimal proportions during the last 30 years. Furthermore, currency overvaluation has encouraged the importation of cheaper foodstuffs which has been an additional deterrent to agricultural production.

Since the discovery of oil at the beginning of this century, imports have been a dominant element in the Venezuelan economy. During the 1960s, Venezuela imported 25 per cent of the total food consumption. In the 1970s, this proportion increased to 70 per cent (Pinto Cohen 1984). In a very optimistic view of the recent history of agriculture in Venezuela, Kastner et al. (1983) blame cereals for the high importation levels and show that most of the imports go to the food industries. This has occurred despite a national strategy that has aimed to replace imports with locally produced products. Development of the agro-industry has not increased agricultural production. Instead, using national subsidies (oil money), they have imported unprocessed agricultural products, taken over crop storage and trading arrangements, and ruined rural producers. This has been possible because the agro-industrial sector is closely tied to large financial corporations and

254

to the commercial sector, operates as a monopoly and has considerable political influence. Also, the agro-industry has been developed using technology purchased from developed countries, which is unsuitable for the quality of national agricultural production (Kastner et al. 1983).

Following the recipes of the International Monetary Fund (IMF) and the World Bank, the present administration has changed from a strategy of subsidising agriculture to one aimed at transforming agriculture into an industry that will be able to compete on the international market.

DISCUSSION AND CONCLUSIONS

The data presented show that savannas in the area considered in this paper are little used, characterised by extensive cattle-raising with limited managerial input and very low productivity. Crop agriculture, both of traditional as well as introduced crops, remains insignificant as a form of land use. Without subsidies little or no agricultural production would occur within the Orinoco savannas.

Venezuelan savannas are marginal to agricultural development and consequently prone to environmental damage. A strategy of use based on extensive cattle-raising, coupled to technically based improvement of the quality of native pastures and the remodelling of the trading system, allow this situation to improve without major environmental risks. This strategy could also play an important role in developing a livestock production system that can continue to compete on international markets. Other economic activities such as tourism and recreation, as well as the exploitation of native animal populations, and the continued extraction of oil reserves, would help to develop these areas. Further increase of savanna crop agriculture, with the replacement of native vegetation by agricultural fields, and the intensive use of fertilisers and pesticides, however, will undoubtedly create major environmental disturbances in this extended region. This will affect not only the vegetation and the soils, but also and especially the fauna and the aboriginal groups of people. The former has several endemic taxa which are already endangered. The latter represent very peculiar savannic groups that persist only in the region of the Orinoco Llanos.

In both aspects of agricultural activities, livestock and crops, these savannas are of secondary importance when compared to other regions of the country, where agriculture has replaced more productive natural systems. Although ecological constraints to productivity limit the agricultural potential of these savannas, it seems clear that their present condition is more the result of complex social, economic and political processes that began with the European occupation of Venezuela five centuries ago.

Since the proportion of savanna land used for crop agriculture is small,

255

and the use of technical support is very moderate, there are no major environmental problems caused by crop production within the Venezuelan savannas. Since regulations for the use of fertilisers and pesticides are not enforced, however, savanna agriculture may be contributing to the pollution of the rivers in the Orinoco basin.

Pollution from the oil industry is local and does not represent any major problem. It has been suggested that annual burning may be a menace for the survival of endangered animal species (MARNR-BIOMA 1991). However, there is no evidence available to support this.

The development of Venezuelan agriculture does not seem to be different from many other tropical underdeveloped countries, despite the abundance of resources provided by oil exports. Technical solutions, such as pasture improvement with low investment technologies, can be found and are currently under scrutiny some have already been experienced successfully (Thomas et al. 1990). But beyond the technical problems it is the socio-political framework which is conditioning development. Pouring money into a system that cannot hope to ultimately compete on international markets is likely to be counter-productive and would only suppress opportunities for economic development elsewhere. Moreover, the experience so far suggests that the usual impact of attempts to improve the productivity of Venezuelan savannas is the further entrenchment of social inequities.

ACKNOWLEDGEMENTS

We thank M. Pino and M.G. Silva for help with the sources. We are also grateful to M.D. Young and O. Solbrig for their suggestions with the first draft of the manuscript. This paper was partially funded by CONICIT grant S1-1968, and by CDCHT-ULA grant C-411-90.

REFERENCES

Acevedo, M.F. and Silva, J.F. (1985) Información ambiental y ecológica sobre los Llanos del río Orinoco. Serie de Informes Técnicos DGSPOA/IT/MARNR. Ministerio del Ambiente y los Recursos Naturales, Caracas. Venezuela, 290 pp.

Berrade, F. and Tejos, R. (1984) Productividad primaria aerea neta en diferentes unidades fisiográficas del módulo "Fernando Corrales", Apure, Venezuela. *Revista UNELLEZ de Ciencia y Tecnología* (Barinas, Venezuela) 2:17–34.

CENDES (1975) Estudio de prediagnóstico para el plan maestro de ordenamiento territorial del area de la faja petrolífera del Orinoco. Fase I: Prediagnóstico de la Sub-área Anzoátegui-Monagas. Centro de Estudios del Desarrollo, Universidad Central de Venezuela, Caracas, 549pp.

Cochrane, T.T., Azevedo, de L.G., Thomas, D., Madeira Netto, J., Adamoli, J. and Verdesio, J.J. (1985) Land use and productive potential of American savannas. In: J.C. Tothill and J.J. Mott (eds), *Ecology and management of the world's savannas*. Australian Academy of Sciences, Canberra, pp. 114–24.

Comerma, J. and Luque, O. (1971) Los principales suelos y paisajes del Estado Apure. *Agronomia Tropical*, 21:379–96.

COPLANARH (1974) Estudio geomorfológico de los llanos orientales, regiones 7 y 8. Comisión del Plan Nacional de Aprovechamiento de los Recursos Hidraúlicos (Coplanarh), Venezuela. Publicación No. 38, 164 pp.

Fariñas, M.R. and San José, J.J. (1985) Cambios en el estrato herbaceo de una parcela protegida del fuego y el pastoreo durante 23 años. *Acta Científica Venezolana*, 36:199–200.

Gil Beroes, R.A. (1976) Producción y manejo de pastos en las sabanas inundables del Alto Apure. *Boletín de la Sociedad Venezolana de Ciencias Naturales*, 32:103–14.

Hernandez, J.L. and Merz, G. (1986) Los cereales en el patrón de consumo de transición, posibilidades de cambio en los próximos años. *Cuadernos del CENDES*, 6:97–123.

Kastner, G., Austin, J. and Tello, M.T. (1983) La Venezuela agrícola, mitos y realidades. Ediciones IESA, Caracas, Venezuela, 73 pp.

Lopez-Hernandez, D., Niño, M., Garcia, L. and Carrion N. (1983) Annual budgets of some elements in a flooded savanna (Módulo Experimental, Mantecal, Venezuela). *Ecological Bulletin* (Stockholm), 35:541–45.

MARNR (1978) Parques Nacionales y Monumentos Naturales de Venezuela. Instituto Nacional de Parques, Ministerio del Ambiente y los Recursos Naturales Renovables, Caracas, 192pp.

MARNR (1982) Estudio agrológico gran visión del Estado Apure. Serie Informes Tecnicos. DGSIIA/IT/110. Ministerio del Ambiente y los Recursos Naturales Renovables, Caracas, Venezuela, 140pp.

MARNR-BIOMA (1991) Plan de Ordenación y Manejo del Parque Nacional Santos Luzardo. Tomo I. Ministerio del Ambiente y los Recursos Naturales Renovables, Caracas, Venezuela, 350pp.

Martinez, M. (1975) Diagnóstico técnico-económico de las explotaciones ganaderas del Estado Monagas. Ministerio de Agricultura y Cría. Proyecto MAC-FAO VEN 17. Maturín, Venezuela, 80pp.

Mérida, J.M. (1966) La cuestión indigena en Venezuela. *Cuadernos de la Corporación Venezolana de Fomento*, 3:49–67.

Orta, C.S. (1974a) Impacto de los ingresos petroleros sobre el crecimiento del sector agrícola. Facultad de Ciencias Económicas y Sociales, Universidad Central de Venezuela, Caracas, Venezuela, 67pp.

Orta, C.S. (1974b) Los obstaculos al crecimiento autosostenido de la agricultura venezolana. In D.F. Maza Zavala, (ed) *Venezuela, crecimiento sin desarrollo*. Ediciones Nuestro Tiempo, Mexico, D.F. pp. 201–37.

Pinto Cohen, G. (1984) La Agricultura: revisión de una leyenda negra. In M. Naim, and R. Piñango (eds) *El caso Venezuela: una ilusión de armonía*. Ediciones IESA, Caracas, Venezuela, pp. 500–35.

Rodriguez Mirabal, A.G. (1987) La formación del latifundio ganadero en los Llanos de Apure: 1750–1800. *Academia Nacional de la Historia*. Caracas, Venezuela, 371pp.

San José, J.J. and Fariñas, M.R. (1983) Changes in tree density and species composition in a protected Trachypogon savanna, Venezuela. *Ecology*, 64:447–53.

Sarmiento, G. (1983) The savannas of tropical America. In: F. Bourliere (ed) *Tropical savannas*. Elsevier, Amsterdam, pp. 245–88.

Sarmiento, G. (1984) The ecology of neotropical savannas. Harvard University Press, 235 pp.

Sarmiento, G., Monasterio, M. and Silva, J.F. (1971) Reconocimiento ecológico de los Llanos Occidentales. IV. El oeste del Estado Apure. *Acta Científica Venezolana*, 22:170–80.

Silva, J.F., Raventos, J. and Caswell, H. (1990) Fire and fire exclusion effects on the growth and survival of two savanna grasses. *Acta Oecologica*, 11:783–800.

Silva, J.F., Raventos, J., Caswell, H. and Trevisan, M.C. (1991) Population responses to fire in a tropical savanna grass: A matrix population approach. *Journal of Ecology*, 79, 345–56.

Tejos, R. (1984) Efecto de la altura de la inundación sobre los cambios de vegetación de la sabana modulada "El Rosero", Apure, Venezuela. *Revista UNELLEZ de Ciencia y Tecnología* (Barinas, Venezuela) 2:57–68.

CHAPTER 13

ECOLOGICAL IMPACT OF AGRICULTURAL DEVELOPMENT IN THE BRAZILIAN CERRADOS

C.A. Klink, A.G. Moreira and O.T. Solbrig

SUMMARY

From 1960 to 1980 government policies stimulated the agricultural development of Central Brazil. The impact has been selective but the social,
economic and environmental costs have been high. Subsidies have favoured
commercial crops at the expense of staple crops, national economic
development and the environment. Concessional loans, available in proportion to crop area (not production), have had a dramatic impact on the
nature of development and the area cleared.

Increases in production per hectare have been relatively small, most
being achieved by bringing more land under crops. Although the extent of
environmental modification caused by these policies is less well documented
than economical and social transformations, it is clear that the net impact
of these changes has been negative. Today, habitat loss, water pollution
and erosion are common in the *cerrado* region.

The impact of many of the agricultural development policy initiatives
pursued since 1960 has also been inequitable. Most of the programmes
have favoured wealthy farmers. Taxation arrangements allow people to
avoid paying tax by investing in agricultural land. For those who can afford
it, land has been an excellent hedge against inflation. Credit concessions
have also provided a strong incentive to open up new land rather than
increase the productivity of existing land. Collectively, all these policies
have acted to reinforce the already uneven distribution of land and wealth.

INTRODUCTION

From its beginning as an independent nation in 1822, Brazil has seen its
future as one that requires the development of its vast interior. Initially,

259

scientists and explorers were engaged to traverse and describe the interior. The German botanist Martius, for example, prepared a flora of Brazil that is a masterpiece of nineteenth century plant science. Largely due to its vastness, however, much of the interior of Brazil remained unexplored, uninhabited and undeveloped until the middle of the twentieth century. This state of affairs began to change after World War II, when the development of the interior became a stated policy of the various governments that ruled Brazil during this period. Massive road and infrastructure construction began but even so, in comparison with Brazil's coastal areas, much of the centre remains relatively undeveloped.

Geopolitically Brazil is divided into five administrative regions: North, North-east, Center-West, South-east and South (see Figure 13.1). Two landscapes dominate the interior of Brazil: the Amazon rainforest, the largest such ecosystem in the world; and the seasonal savannas, known locally as the *cerrado*. The *cerrado* is located largely within the Center-West region. Throughout the rest of this paper we refer to this region as Central Brazil.

THE CERRADO

The word "*cerrado*", in its wider sense, designates several structural types of vegetation, from dense pure grasslands to woodlands with a fairly continuous tree canopy (Eiten 1972; Coutinho 1978). A quantitative analysis of *cerrado* vegetation shows a variation in physiognomy and species composition, and a significant correlation has been established between vegetation structural gradients and nutrient levels in the soil in certain areas (Goodland and Pollard 1973).

Climate

The total annual rainfall in the *cerrado* region varies from 1000 to 1800 mm. Its distribution is characterised by distinct wet and dry seasons, with 80 per cent of the total precipitation falling between November and April. In short, this tropical hot subhumid climate has a rainy summer and a dry winter.

The length of the dry season varies within the region from about 3–6 months. Water stress in the topsoil during this season is a major constraint to crop production. Most of the crop production in the region occurs during the wet season, starting with soil preparation in October–November, followed by planting in December and harvesting in April–May. A second harvest is possible only in irrigated areas (EMBRAPA/CPAC 1981).

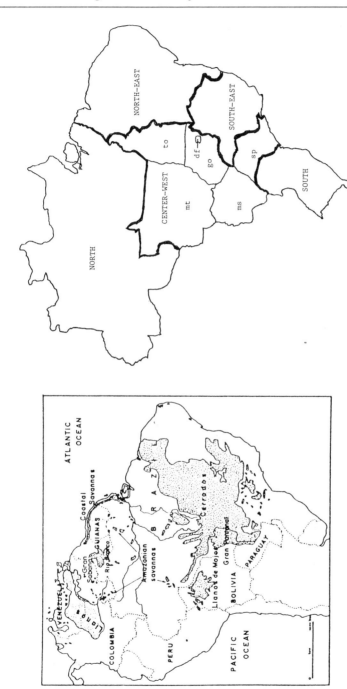

Figure 13.1 Major savannas of South America, showing the *cerrado* region in Central Brazil (from Sarmiento 1983) and the five Brazilian administrative regions: North, North-East, Center West (Central Brazil), South-East, and South. The Center-West region contains the states of Mato Grosso (mt), Mato Grosso do Sul (ms), Goiás (go), Tocantins (to), and the Federal District (Brasilia)

During the wet season, short-term droughts or erratic periods of mid wet-season water stress, called *veranicos* may seriously affect production. According to Cochrane et al. (1988), 60 per cent of the *cerrados* are affected to some extent by this phenomenon, and 28 per cent is quite susceptible. When a short-term drought coincides with a crucial stage of crop growth, such as flowering, drastic yield losses can occur.

Soil

Some 50 per cent of the *cerrado* is covered with oxisols (Goedert 1983). Generally their physical properties – depth, micro-aggregate stability, drainage, slopes commonly less than 3 per cent – favour agricultural mechanisation.

The water-holding capacity of *cerrado's* soils is very low. For example, only 6–8 per cent of the water is retained at tensions of less than one bar in a dark red latosol (typic Haplustox) at 0–45 cm depth (Garrido et al. 1979). This means that the water stored at the plough layer is enough to maintain a crop in full growth for only 6–10 days (Goedert et al. 1980).

The nutritional condition of most *cerrado* soils is poor and, under natural conditions, productivity is low. Exchangeable calcium and magnesium are below critical levels for plant nutrition. Phosphorus is the most deficient plant nutrient and, because the clay in most oxisols has a high phosphate absorption capacity, successful crop production requires large amounts of phosphate fertiliser. The cation exchange capacity is low and often saturated with aluminium, which may attain toxic levels for most crops (Lopes and Cox 1977). The soils are also very acidic. However, once phosphate fertiliser and lime are applied to correct for both the nutrient deficiency and acidity, soils become productive. Fortunately, Brazil has large reserves of both these minerals (Goedert et al. 1980).

Flora and fauna

The *cerrado* region has one of the richest savanna floras in the world, especially of woody species. Every small area shows a rich and diversified flora. According to Heringer et al. (1977), of the 774 woody species, 429 compose the proper floristic stock of the *cerrados*, 300 species belong to forest formations and the remaining 45 to other vegetation types. This number of 429 unique woody savanna species is not matched by any other savanna flora in the world. When compared with other savannas, the herbaceous flora, including the grasses, however, is not as rich as the woody flora (Sarmiento 1983).

Considering its geographical extension, the mammalian fauna of the *cerrado* region is rather poor. Most of them move between the savanna and other surrounding ecosystems. Around 160 species have been recorded (Ojasti 1990). Although most of the region is covered with seasonal savannas, the majority of species is associated with or restricted to forested patches, as for example the gallery forests (Sick 1965).

Most large herbivores are ungulates. The scarcity of these animals today contrasts with the rich fauna of the Late Cretaceous, where it attained its maximum diversity, similar to the present African ungulate fauna (Ojasti 1983).

When compared to African savannas, the *cerrado* avifauna is meagre. Most species are arboreal. A few characteristic large land birds such as the rheas, screamers and the seriema (*Cariama cristata*) are endemic to the region (Fry 1983).

The invertebrate fauna is poorly known. The most abundant groups are insects and spiders. Agricultural development brought large areas under mono-culture and introduced some insect-pests to the region, such as the "cigarrinha-das-pastagens", an homopthera (*Deois flavopicta*) that severely attacks cultivated pastures (Cosenza 1988).

LAND USE IN THE CERRADO

Since the 1960s, Central Brazil has been portrayed as a transition zone between the developed South-east and the unoccupied North of the country. Today, many people still see it as as an agricultural frontier linked to the nation's economic and production systems. Its main agricultural crops are rice, soybeans, maize, beans and cassava. Further agricultural development of the region is presented as a means to advance capitalistic objectives and enhance the national economy (Duarte 1989). Politically, it encompasses the states of Mato Grosso, Mato Grosso do Sul (formerly southern Mato Grosso, split from it in 1979), Goiás, Tocantins (formerly northern Goiás, split from it in 1989), and the Federal District (Brasilia).

Until 30 years ago the *cerrado* of Central Brazil was used primarily for extensive cattle-raising. Today, it is estimated that 37 per cent of its natural vegetation has been transformed into cultivated pasture, crops, dams, urban settlements, and degraded areas (mainly due to mining, erosion, abandoned farms) (Dias 1990).

The remaining area can be divided into two categories: (1) "Managed" natural areas (56 per cent of the total), as, for example, native pasture for cattle raising, forests for either timber exploitation or protection of watersheds, charcoal production, environmental protection areas, indigenous reservations, and military training areas; (2) natural reserves (6.6 per

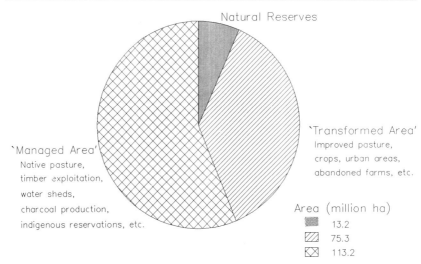

Figure 13.2 Estimated current pattern of land use in the *cerrado* (Source: Dias (1990))

cent of the total), including parks, natural monuments, habitats for wildlife protection, and scientific reservations (see Figure 13.2).

Cattle-raising

Brazil has one of the world's largest cattle herds, approximately 140 million head. Extensive cattle-raising has been the most common land use of the savannas. Grasses are abundant during the rainy season but they dry out during the dry season (see Chapter 2 in this book). During the dry season and wherever it is possible, cattle graze on the edges of streams and marshes where some green pick remains. Generally, farmers burn the native pasture at the end of the dry season to induce regrowth for grazing. Due to the scarcity, unpalatability and low protein and phosphate content of the pasture during the dry season, however, beef cattle production per head is generally poor. Stocking rates average 0.2 animal units per ha per year, annual reproductive rates are between 35 and 45 per cent and annual weight gains are around 20 kg/ha per year (Eiten and Goodland 1979; Goedert 1983).

Most Brazilian herds are the result of a cross between Zebu (*Bos indicus*) and European domestic breeds (*Bos taurus*). The Zebu variety's heat and tick-resistance is advantageous on open range, but the European breeds have a greater reproduction rate (Eiten and Goodland 1979). A cross between the two breeds is used to combine the climatic advantages of the Zebu with the greater productivity of the European.

264

Table 13.1 Number of cattle in Central Brazil from 1965 to 1987. Source: IBGE (1960–1987)

Year	Number (× 1000)	Share of Brazilian herd (%)
1965	20,794	23
1970	20,048	20
1975	27,522	19
1980	33,673	28
1985	41,126	32
1987	43,962	32

The establishment of cultivated pasture has been traditional in the forested areas of the region, but during the 1970s pasture cultivation was extended to the savanna areas. As a result, the area under cultivated pasture in Central Brazil increased by 15.6 million ha (Mesquita 1989). A common practice is to clear-cut and burn the savannas, and then seed with various grasses, generally of African origin as, for example, species of *Andropogon*, *Brachiaria*, *Hyparrhenia*, and *Melinis*. Legume species, like *Centrosema* and *Stylosanthes*, are starting to be used as source of protein (Eiten and Goodland 1979). Changing from the traditional extensive use of the natural savannas to cultivated pastures has increased the region's share of the national herd (see Table 13.1). The establishment of cultivated pastures has opened new possibilities for cattle-raising, since it allows fattening of cattle in the region itself, without the necessity of taking them to São Paulo or Minas Gerais states for fattening (Mesquita 1989).

Crop production

In a technical sense, the savannas of Central Brazil have always been considered appropriate for agriculture. It is estimated that, using modern techniques, at least 50 million ha could be developed for crop production (Wagner 1981).

Substantial growth in the production of maize, rice, soybean and wheat has occurred, whilst cassava and bean output has remained unchanged from 1960 to 1987 (see Figure 13.3).

The most dramatic increase has occurred in soybean production. Virtually non-existent as a crop in 1960, some 5.8 million tonnes were grown in 1987.

Sugar cane production is concentrated in the state of Goiás, and has benefited from the national programme that subsidises alcohol production.

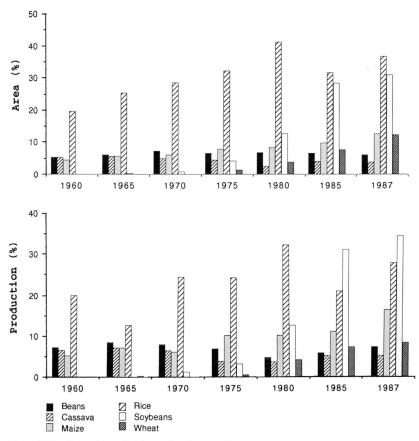

Figure 13.3 Central Brazil's share of national agricultural production by principal crop type. Source: IBGE (1960–1987)

Production increased from 297,000 tonnes in 1970 to 1.3 million tonnes in 1980 (Mesquita 1989).

The number of tractors has increased more than six times in the same period (Gusmão 1988). Industrial inputs now represent one-quarter of total farming costs. According to Goedert and Lobato (1987) the total cost of transforming the *cerrado* is US$ 800 per ha, and 42 per cent (US$ 340) of this total comes from inputs such as lime and fertilisers.

To understand how agriculture and commercial livestock production have evolved in Central Brazil, however, it is necessary to understand the history of Brazil, the interdependence of its various regions and the way that government programmes have manipulated economic incentives.

ECONOMY AND AGRICULTURE IN CENTRAL BRAZIL
BEFORE 1960

Exploration of southern Mato Grosso started in the middle of the nineteenth century after the Paraguay War. Concerned about its very low population density and its being a border province, the imperial authorities decided to stimulate the occupation of this region after the war by providing economic incentives for the growth of tea in Dourados and Ponta Porã (Almeida and Lima 1956). Because of the existence of large areas of savannas, cattle-raising began on a few large ranches in Campo Grande and Corumbá (Mueller and Penna 1978).

Occupation of Goiás is more recent and has been more intense than in Mato Grosso. This late development was probably due to the scarcity of road communication within Goiás and the lack of interest of the central government, since the State is not a border one.

In the 1920–1930 period, the population of São Paulo, in the South-east, expanded. São Paulo became Brazil's wealthiest State. Its economy was based upon strong coffee growing and processing industries. The capital city (of the same name) has become a main market for cattle from Goiás and Mato Grosso (Almeida and Lima 1956; Hees et al. 1987).

Goiás also benefited from its proximity to São Paulo, and the State saw a growth of its agricultural activities during the 1930s. Significantly, the Getulio Vargas government (1930–1945), undertook a colonisation policy in Goiás state – the "marcha para o oeste" – that encouraged people to settle on fertile forested land, which was more suited to agriculture development than ranching (Hees et al. 1987; Mueller and Penna 1978). As a result of this colonisation process, land tenure distribution in Goiás is less inequitable than in Mato Grosso (Mueller and Penna 1978).

Infrastructure

The lack of a river system linking the region with the urban nuclei of the coast has always been an obstacle to the development of Central Brazil. Railroad construction started in the beginning of the twentieth century but went very slowly. The first railway connection to Mato Grosso was initiated in 1905, linking São Paulo to southern Mato Grosso (Lucarelli et al. 1989). It reached Campo Grande in 1914 and brought a substantial economic impulse to the city (Mueller and Penna 1978). The railroad was finished in 1947 when it reached the Paraguay river (Lucarelli et al. 1989). Another railway linking São Paulo with Anápolis in Goiás was completed in 1935 (Almeida and Lima 1956).

After 1946, the federal government promoted the construction of roads

Table 13.2 The population of Central
Brazil from 1872 to 1980. Source: Ajara
(1989)

Year	Population	Density (*n* per km^2)
1872	220,812	0.12
1890	320,399	0.17
1900	373,309	0.20
1920	758,531	0.40
1940	1,258,679	0.67
1950	1,736,965	0.92
1960	3,006,866	1.59
1970	5,167,203	2.72
1980	7,544,795	4.02

as the means to link the Brazilian regions (Lucarelli et al. 1989). Once the decision to build a new capital, Brasilia, was taken in the late 1950s, the Belo Horizonte-Brasilia road (paved) to the South-east, the Belém-Brasilia road (unpaved) to the North and a connection to São Paulo were constructed.

Population growth and urbanisation

Total population in Central Brazil grew by 73 per cent between 1950 and 1960. Most of this influx, however, was due to the employment opportunities associated with the construction of Brazil's new capital, Brasilia (Lucarelli et al. 1989). Nevertheless, in the hundred years before 1960, the regional population grew at a rate twice that of the rest of Brazil (see Table 13.2). The growth was, in part, due to internal migration (Ajara 1989).

Throughout Central Brazil there has been a strong trend towards urbanisation since the 1940s (Table 13.3). While the rural population increased by about 56 per cent from 1950 to 1960, the urban population almost tripled for the same period, with important implications for the rate at which agricultural colonisation has proceeded.

Agriculture

Until the late 1950s, the contribution of Central Brazil to the nation's agricultural output was very low. Both the extent of land under crops and total production contributed less than 10 per cent to the national total for all crops except rice (see Table 13.4). Although cattle-raising is conducted

Table 13.3 The total urban and rural population of Central Brazil from 1940 to 1980. Source: Ajara (1989)

Year	Rural	Urban*	Per cent rural
1940	987,842	270,837	78
1950	1,313,468	423,497	76
1960	1,953,760	1,105,150	64
1970	2,674,192	2,493,011	52
1980	2,430,198	5,114,597	32

* Brazilian statistics do not indicate the cut-off population density for a town.

Table 13.4 Planted area, production, and the contribution of Central Brazil to the total Brazilian production of some crops in 1960. Source: IBGE (1960–1987)

Crop	Area (ha)	Production (tonnes)	Central Brazil share Area (%)	Central Brazil share Production (%)
Bean	132,769	126,032	5.2	7.3
Cassava	69,015	1,170,852	5.1	6.7
Coffee	86,363	118,839	1.9	2.8
Cotton	17,177	19,960	0.6	1.2
Maize	285,578	434,087	4.3	5.0
Rice	577,081	945,303	19.5	19.7
Sugar cane	45,557	1,964,217	3.4	3.4
Wheat	349	414	0.03	0.06

on an extensive basis, the cattle herd comprised only one-fifth of the national herd.

DEVELOPMENT OF AGRICULTURE IN CENTRAL BRAZIL FROM 1960 TO 1980

The overall objective of policy planners in Brazil after World War II was to develop local industry by restricting imports. This was implemented by subjecting imported goods to heavy import duties that often exceeded 100 per cent and by overvaluing the cruzeiro. The overvalued cruzeiro acted as a disincentive to the production of crops for export. As a direct result, exports of coffee and sugar declined. Simultaneously, in the face of inflation, the government set food price ceilings to favour the urban population. Low food prices acted as a further disincentive to farm producers (World Bank 1982).

As would be expected from such a regime, agricultural production did not keep pace with overall growth in the economy and the importation of food became necessary (Soskin 1988). Ultimately, a series of poor harvests, combined with the lack of incentives, led to a sudden rise in food prices in the late 1950s and a severe supply crisis in the early 1960s (World Bank 1982). The rise in food prices pushed up inflation, causing social and political unrest, and eventually led to the collapse of the democratic government in 1964.

Attempting to repair the damage, the military government of General Arthur da Costa e Silva (1967–1969) made agricultural development a key priority, for three reasons. First, the population of Brazil was growing, especially the urban sector, and it was perceived that, if agriculture could not keep up with demand, food price increases would add to inflation, which the government was trying to bring under control. Second, increased agricultural exports were seen as one of the fastest ways to supply the foreign exchange needed to buy equipment for the expansion of the highly protected secondary industry. Third, increasing rural incomes was seen as necessary to stem the rural population exodus that was becoming a serious problem.

The new regime introduced a programme that aimed to modernise agriculture. The federal government waived taxes on farm profits, locally produced tractors, farm machinery, pesticides and fertilisers (Hees et al. 1987). These actions stimulated the export sector enormously. Soybeans went from an almost unknown crop in 1960 to the country's largest.

At the same time, the minimum-price programme was expanded, and today covers 42 commodities ranging from products of national importance, such as black beans, rice and corn, to localised items like *guaraná* (World Bank 1982). With assurance from a government-guaranteed minimum price, some uncertainty was removed and farmers had a greater incentive to maintain or increase their investment. Guaranteed minimum prices also protected farmers from the effects of price control policies that aimed to reduce the cost of food in urban areas.

Infrastructure

In the 1970s the Belém-Brasilia road was paved and other road connections to the Amazon and to the North-east were built. Today, the majority of State capitals and main economic centres are connected to Brasilia. Parallel to the construction of roads, a railway connection was established from Brasilia to São Paulo and Porto Alegre in the South. The government also built secondary roads and enlarged the rural electric network (Duarte 1989;

Table 13.5 Urban population as percentage of the total population for Central Brazil and Brazil from 1960 to 1980. Source: Cardoso (1989)

	1960	*1970*	*1980*
Central Brazil	35	48	67
Brazil	45	56	67

Hees et al. 1987; Mueller and Penna 1978; Scolari 1986). All this growth in infrastructure has acted as a major factor in stimulating agricultural development within Brazil (Southgate 1991).

Population growth and urbanisation

Between 1960 and 1970 the population of Central Brazil grew at roughly double the rate for the Brazilian population as a whole. Much of this growth can be explained by the decision to construct the new capital city Brasilia which, in turn, created many employment opportunities.

A significant proportion of this growth has been the result of internal migration from other parts of Brazil. Many of the people displaced by agricultural land concentration and mechanisation in the South and Southeast of Brazil moved into Central Brazil during this period (Mahar 1989; Southgate 1991). Collectively, the policies adopted by the Brazilian government forced the substantial migration of displaced rural people to urban areas. It is significant that the rural population declined from 1960 to 1980 (see Table 13.5) (Cardoso 1989; Hees et al. 1987).

Credit policy

Credit was one of the principal instruments utilised by the Brazilian government to develop agriculture. Probably nowhere are the environmental, economic and social contradictions of intervention in the economic process better seen than in the credit policy of Brazil in the 1960–1980 period.

The greatest incentive was provided by a policy of low-interest farm loans for periods of 6 months to 12 years. Additionally, because these loans were at fixed rates of 13–15 per cent, they were further subsidised by not taking inflation (running at an average of 40 per cent during this time) into account. Even lower rates were charged for fertilisers and other farm inputs. As a result, credit to the rural sector grew between 1969 and 1979 at a rate

Table 13.6 Distribution of production credits by crop (%) in Brazil for the 1975–1979 period. Source: World Bank (1982)

Crop	1975	1976	1977	1978	1979
Soybeans	18	20	20	20	21
Rice	18	16	16	13	14
Coffee	10	11	13	12	13
Wheat	13	13	11	11	10
Maize	11	11	8	9	10
Sugar cane	11	10	9	9	7
Cotton	5	7	8	6	6
Beans	1	2	3	3	3
Cocoa	1	1	1	2	1
Cassava	0	1	1	1	2
Other	10	9	11	14	13
Total*	100	100	100	100	100

* Excludes non crop-specific credits

188 per cent faster than total output (199 per cent for agriculture, 164 per cent for livestock production). The growth was particularly high during the 1969–1976 period when agricultural credit rose at a yearly 24 per cent compound interest rate (World Bank 1982). After 1977 credit flows slowed down, especially long-term investment credit.

About half of the loans (45 per cent) were short-term production loans, either due when the crops were brought in or, alternatively, repayable in 6 months to 2 years according to activity. Approximately a third of the loans were long term and payable in 5–12 years. The rest of the loans were marketing credits largely associated with the minimum-price programme. Loans were not evenly distributed by crops. Over 75 per cent of them were concentrated in six crops: soybeans, rice, coffee, wheat, maize and sugarcane (see Table 13.6). Soybeans alone received one-fifth of the credit made available to Brazilian farmers.

The very rapid growth of credit in relation to output raises doubts about the effectiveness of using concessional loans as a way to increase production, and highlights the degree to which agricultural policies have provided opportunities for those who can afford to purchase farm land to increase their wealth. Furthermore, since loans were allocated on the basis of area planted, they encouraged extensive and inefficient agriculture (Goedert 1990). Credit was concentrated in the South and South-east regions, which received almost twice as much credit per ha as the savanna region (Central Brazil), which in turn received 70 per cent more than the North and North-east (where the small producers of food crops are concentrated). This, in

Table 13.7 Expansion of the area of rural establishments in Central Brazil from 1940 to 1980. Source: Gusmão (1988)

Years	Total area (million ha)	Increase (%)
1940	40.3	n.a.
1950	53.6	33
1960	79.9	49
1970	81.7	2
1980	113.4	39

turn, drove up land prices, and increased demand in the *cerrado* where land, because of its lower productivity, was cheaper (Southgate 1991).

Land ownership, income tax and investment

All of the above adverse policy impacts were further compounded by income tax laws. Under Brazilian income tax law, practically all profits earned or diverted through an agricultural enterprise were tax exempt (Binswanger 1989). Thus, a businessman who incurred a tax liability by running a profitable urban business could buy a farm and use tax savings to make a profit.

The tendency for land to be owned by wealthy people has been further aggravated by Brazil's high inflation rate. The average rate has been around 40 per cent per annum, with the consequence that buying a farm has been an excellent way to hedge against inflation (Mueller and Penna 1978).

As a consequence of these policies there has been a significant increase in the area officially under agricultural and livestock production (see Table 13.7). In Aripuanã and Alta Floresta (northern Mato Grosso) and during the 10 years between 1970 and 1980, for example, whilst the area under effective production stayed the same, the occupied area increased by more than 838,000 ha (Mesquita 1989).

All the above conditions have caused land ownership to become highly concentrated. In Central Brazil, 74 per cent of the land area was owned by 7.5 per cent of "farmers" in 1980 (see Figures 13.4 and 13.5). Large areas were underutilised and those that were utilised tended to either support more mechanised forms of agriculture or cattle-raising as their main activity (Hees et al. 1987). Most of the labour was employed on small holdings. Nearly 50 per cent of rural labour was employed on farms less than 100 ha in area.

In some areas a considerable proportion of land is owned by absentee

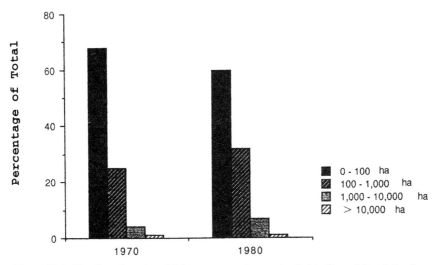

Figure 13.4 Number of rural establishments according to size (ha) in Central Brazil, for the 1970–1980 period. Source: Hees et al. 1987

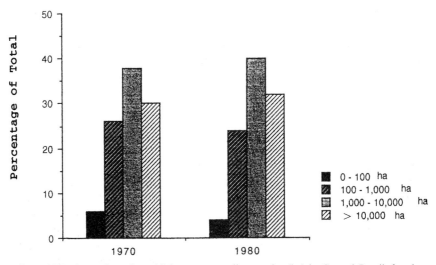

Figure 13.5 Area of rural establishments according to size (ha) in Central Brazil, for the 1970–1980 period. Source: Hees et al. 1987

landlords. In the Bico de Papagaio of northern Goiás (currently Tocantins), for example, 80 per cent of farmers lease their land from other people. Many of these people aspire to own their own land and the resulting social tension has led to the invasion of public and private land (Mesquita 1989).

Table 13.8 Soil losses in tonnes per ha under different cropping systems on a clayey Dark Red Latosol (Haplustox) with a 5.5% slope under natural rainfall. Source: Dedecek et al. (1986)

Years	Fallow	Maize	Rice	Soybeans*	Soybeans (no tillage)	Natural vegetation	Rainfall (mm)
79/80	183	87	39	13	2	0.2	1661
80/81	5	2	1	1	0.5	0.1	1172
81/82	13	13	4	7	4	0.1	1683
82/83	25	23	3	8	5	0.1	1644
83/84	38	20	16	11	6	0.1	1669
84/85	54	28	10	10	8	0.2	1830

*Conventional system

ECOLOGICAL IMPACT OF AGRICULTURAL DEVELOPMENT

Ecological impact was not a major concern when the policies to favour agriculture were implemented in the late 1960s. There has been little awareness of the environmental consequences of agriculture and livestock-raising in Central Brazil. However, the transformation of the natural ecosystem has caused changes in soil structure, surface water flow, vegetation cover and wildlife abundance in the *cerrados*.

Soil

In terms of soil conservation agricultural development has had some negative consequences. Despite only a few years of intensive cultivation, large areas are now facing problems of soil compaction and erosion. Some authors (Dedecek 1986; Morais 1985) attribute these problems to environmentally inappropriate management practices, such as mechanised fallow at the end of the dry season and leaving soil bare exactly when the probability of localised thunderstorms is highest. Soil losses on a 5 per cent slope under bare fallow can be as high as 130 tonnes/ha per year, but can be decreased to less than 2 tonnes/ha per year if the soil surface is covered with a soybean crop established under a no-tillage system (Goedert et al. 1982).

Dedecek et al. (1986) evaluated soil and nutrient losses of a *cerrado* soil under different cropping systems. Some of their results are shown in Table 13.8. Generally, they show that reduction in soil loss is directly correlated with the amount of plant cover. Maize (*Zea mays*), due to its architecture and planting interval (1 m), was the most erosive crop. Soybeans, planted in smaller intervals (0.5 m) and with more biomass coverage closer to the ground, was the least erosive annual crop, and no-tillage soybeans showed an even greater reduction in soil losses.

As many researchers have concluded, any programme of erosion control must address soil cover. The major risks occur during and immediately after crop sowing. The impact of raindrops can be effectively minimised by no-tillage or minimum-tillage operations. A common practice in the region is to pile the vegetation collected during clearing operations into contour bands, spaced according to the slope, which leads to the building of a terrace. Combined with contour seeding, this practice has been acknowledged to be very useful in reducing run-off and, hence, soil erosion.

Nutrient loss is directly related to soil loss. In the low-nutrient *cerrado* soils, most nutrients lost in run-offs are those that have been added in the process of soil preparation for cultivation. With poor soil conservation practices, farmers are losing not only the original soil but the capital invested in lime and fertilisers.

Water

In spite of the relative youthfulness of intensive farming systems in the *cerrado* region, problems of water pollution are already emerging (Souza 1990). Recent studies, for example, have detected nitrogen and pesticide pollution in almost all underground aquifers around Brasilia (J.G. Barros, personal communication). The use of pesticides started to grow in the 1980s, and is closely related to the expansion of cash crop production, such as soybeans. In 1984, Central Brazil expended 9.5 per cent of all pesticides used in Brazil (Gusmão 1988).

Flora

The *cerrado* vegetation is adapted to a low but regular degree of periodic burning. According to Eiten and Goodland (1979), if Brazilian savannas are burned at a 3-year interval, the physiognomy of the *cerrado* is barely disturbed, the herbaceous layer returns to full density within a year, and thin-stemmed shrubs return to full density in a few years.

Population increases and agricultural expansion have led to an increase in burning rates, and large areas still covered with natural vegetation are now burned almost every year. The burning rate in areas still occupied by Indians varies from one burning every 3 years, to one every 10 years, depending on how close the area is to hunting routes (Eiten and Goodland 1979). More frequent burning kills thick-stemmed shrubs and trees and, since these are slower-growing than herbs and thin-stemmed shrubs, the *cerrado* as a whole becomes lower and more open. The effect of these shifts in vegetation structure on the *cerrado* biodiversity is still unknown.

Since agricultural expansion started in more fertile lands, most of the dry forest of southern Goiás and Mato Grosso has disappeared, and some other types of vegetation, such as the mesotrophic woodland, are now becoming rare (Ratter 1971).

Fauna

The impact of agricultural development on large animal populations is not well documented, but it is reasonable to assume that, as agricultural development has disturbed large areas of natural habitat, the impact has been significant. The clearing of large areas for agriculture and cattle-raising, as well as the diseases carried by domestic animals, have reduced many populations. For example, once widely distributed throughout the open *cerrado*, the pampas deer (*Ozotoceros bezoarticus*) is now reduced to a few isolated relict populations (Carvalho 1973).

CONCLUSIONS

Brazil has had an overall desire to populate and develop its vast interior that can be traced to the early days of the Brazilian empire. Until recently, however, no specific policy was successful in this respect. During the democratic government of President Juscelino Kubitschek, the capital was moved from Rio de Janeiro on the coast to Brasilia in the very centre of the country. The military government of General Emilio Medici enunciated a specific policy of opening the interior to colonisation. This was followed by a programme of road-building, of which the Transamazonica is the best-known example. Agricultural colonisation was largely a failure in its stated objective (Smith 1982).

The economic and agricultural policies pursued by the government of Brazil after 1964 were designed to benefit the country as a whole. But the social dislocations created, especially in the southern part of the country, had a strong impact on the settlement patterns in Amazonia and the *cerrado*. More recently, both the government and the public in Brazil have become aware of the environmental consequences of these policies.

Agriculture, and to a lesser extent livestock-raising, always create environmental modification, which in the case of high-input agriculture can be very drastic. As long as it does not create irreversible changes in the capacity of the landscape to sustain humans and their domestic animals, change *per se* is not necessarily negative. The question to ask then is whether the changes that have occurred have reduced the carrying capacity of the *cerrado* and whether these are a direct result of the economic policies pursued by the Brazilian government.

Faced with a severe foreign payment crisis that was in part due to an inability of agricultural production (and exports) to keep pace with increasing demand, between 1954 and 1966, and the corresponding increase in agricultural imports, the government resolved to institute policies that would favour increased agricultural production. The main policies were an overvalued exchange rate, barriers against imported manufactures, concessional credit, control of food prices, and taxation of unprocessed farm products.

Because of these policies the area of cultivated land increased dramatically and came to be dominated by soybean production. Brazil is now the world's largest exporter of soybean products, but although soybeans have displaced coffee, they have not yet acquired the prominence in the economic life of the country that sugar, rubber or coffee had in the past. Most of Brazil's export-earning capacity now comes from the production of manufactured goods. In addition, agricultural exports are more diversified.

The increase in agricultural output took place in an economic environment that was highly distorted by government policies. On the one hand, the overvalued cruzeiro during much of the 1960–1980 period made imported foods relatively cheap. At the same time, however, the subsidised price of certain staples, especially wheat products, acted as a disincentive to their production. To counteract this negative effect (as a part of the industrial development policies) the government instituted a policy of subsidised credits and minimum crop prices. We have no indicators to critically evaluate the inefficiencies introduced by these economic distortions, but all indications are that they were substantial, and that they significantly reduced opportunities for economic growth (World Bank 1982). Production decisions were increasingly made during this period based on expectations of inflation (which determined the actual credit subsidy of fixed interest loans and the difference between minimum prices and market prices) rather than on economic production expectations. Furthermore, since credits were based on area planted rather than on land productivity, the credit policies favoured extensive cultivation. Land tenure policies had a similar effect.

Another government policy favouring agriculture was the subsidisation of fertilisers. High-yielding commercial agriculture in the *cerrado* is not possible without a heavy investment in soil preparation, that includes application of lime and phosphate and nitrogen fertiliser. Because the *cerrado* region requires more fertiliser than other regions, it was directly favoured by the availability of subsidised credit (zero nominal interest) for the purchase of fertiliser, as well as direct price subsidies paid to the farmer. Almost three-quarters of fertiliser subsidies went to the same six crops favoured by credit loans. And since fertilisers are used almost exclusively by medium and large farmers, this subsidy was very inequitable.

Further biases were introduced by the agricultural research policy, although that was not its intention. The government created EMBRAPA (Empresa Brasileira de Pesquisa Agropecuária) in 1972 to develop agricultural technologies suited to Brazil. The CPAC (Centro de Pesquisa Agropecuária dos cerrados) had as its mission the development of technologies suited to the Brazilian savanna. The performance of EMBRAPA and CPAC have been impressive in almost every respect (Pastore and Alves 1984). However, the technologies that were developed were strongly biased toward medium and large, capital-intensive producers and cash crops (especially soybeans, wheat, maize and rice). Furthermore, their implementation required a degree of education beyond that of the average small producer. The situation was similar in the livestock-raising sector.

The evidence is overwhelming that the agricultural policies of the Brazilian government have been biased in favour of large-scale farmers. This might not have been deliberate but the effect has been the same.

Agricultural credit and accompanying subsidies during the 1960–1980 period probably went to less than 25 per cent of Brazil's farmers (World Bank 1982). The majority of these credits went to only six commodities: soybeans, wheat, rice, maize, coffee and sugar cane. Wheat and rice received credit allocations well in excess of their contribution to total agricultural value. On the other hand, the staple food crops, especially cassava and beans, have been greatly below their contribution to output. Farmers engaged in growing food staples tend to be small farmers, scattered all over the country but concentrated in the North-east region which received much less credit than the South and the South-east, the centres of soybean and wheat production.

RECOMMENDATIONS

Agricultural production in Central Brazil needs to be reassessed in terms of local limiting factors, economic costs, and social and environmental impacts. Policies should promote locally adapted and sustainable agriculture based upon a large effort to conserve basic resources such as water, energy, soil, and plant and animal varieties. Here we suggest some measures that can be implemented to address some of these issues:

(1) There is a need to dismantle the complex web of interacting and conflicting subsidy, tax and credit policies that drive Brazilian agriculture.

(2) Land tenure policies have proven to be socially inequitable and should be changed. Policies directed to agricultural development should give equal emphasis to both production and social equitability.

(3) Development approaches should integrate the small producer more fully into modern agriculture. The goal should be to make the small producer more genuinely self-sufficient. Concessional credit ought to be directed to these farmers.

(4) Research should be carried out in order to establish the particular characteristics of each subregion. Data on natural resources should be collected, organised and synthesised in such a way as to be locally useful. A special effort should be made to integrate scientific with traditional knowledge. Further expansion of the agricultural frontier should only be permitted after more detailed resource data becomes available.

(5) Technologies such as integrated pest management, intercropping and crop rotation should be developed and implemented in order to reduce the use of pesticides and fertilisers.

(6) Among the world's savannas, the *cerrado* vegetation has the highest diversity in terms of number of species. Natural reserves should be created for several purposes, such as maintenance of the biodiversity, protection of wildlife habitats, recreation, and indigenous land.

ACKNOWLEDGEMENTS

We thank C. Cavalcanti, S. Davies, and M.D. Young for their comments on earlier drafts of this paper. C.A. Klink and A.G. Moreira gratefully acknowledge the financial support received from the Conselho Nacional de Desenvolvimento Científico e Tecnológico (CNPq - Brazil).

REFERENCES

Ajara, C. (1989) População. In: A.C. Duarte (ed) *Geografia do Brasil, região Centro-Oeste* (v.1). Fundação Instituto Brasileiro de Geografia e Estatística, Rio de Janeiro, pp.123–48.
Almeida, F.F.M. and Lima, M.A. (1956) The west central plateau and Mato Grosso "Pantanal". 18th International Geographical Congress Excursion Guidebook no. 1. Fundação Instituto Brasileiro de Geografia e Estatística, Rio de Janeiro.
Binswanger, H.P. (1989) Brazilian policies that encourage deforestation in the Amazon. In: World Bank Environment Department Working Paper, no. 16. 24 pp. World Bank Publications, Washington DC.
Cardoso, M.F.T.C. (1989) Organização urbana. In: A.C. Duarte (ed) *Geografia do Brasil, região Centro-Oeste* (v.1). Fundação Instituto Brasileiro de Geografia e Estatística, Rio de Janeiro, pp.189–237.

Carvalho, C.T. (1973) O veado campeiro (Mammalia, Cervidae) – situação e distribuição. *Boletim Técnico da Secretaria de Agricultura de São Paulo* 7:9–26.

Cochrane, T.T., Porras, J.A. and Henão, M.R. (1988) The relative tendency of the Cerrados to be affected by veranicos: A provisional assessment. In: EMBRAPA/CPAC (ed) *VI Simpósio sobre o Cerrado. Savanas: alimento e energia.* EMBRAPA, Brasilia, pp. 229–39.

Cosenza, G.W. (1988) Resistência de gramíneas forrageiras à cigarrinha-das-pastagens, Deois flavopicta (Stal, 1854). In: EMBRAPA/CPAC (ed) *VI Simpósio sobre o Cerrado. Savannas: alimento e energia.* EMBRAPA, Brasilia, pp.507–19.

Coutinho, L.M. (1978) O conceito de Cerrado. *Revista Brasileira de Botânica* 1:17–23.

Dedecek, R.A. (1986) Erosão e práticas conservacionistas nos Cerrados. Circular Técnica da EMBRAPA/CPAC, No. 22.

Dedecek, R.A., Resck, D.V.S. and Freitas Jr, E. (1986) Perdas de solo, água, e nutrientes por erosão em latossolo vermelho-escuro dos Cerrados em diferentes cultivos sob chuva natural. *Revista Brasileira de Ciência do Solo* 10:265–72.

Dias, B.F.S. (1990) Conservação da natureza no Cerrado brasileiro. In: M.N. Pinto (ed) *Cerrado: Caracterização, ocupação e perspectivas.* Brasilia University Press, Brasilia, pp. 583–640.

Duarte, A.C. (1989) Estrutura do espaço regional. In: A.C. Duarte (ed) *Geografia do Brasil, região Centro-Oeste* (v.1). Fundação Instituto Brasileiro de Geografia e Estatística, Rio de Janeiro, pp. 243–66.

Eiten, G. (1972) The Cerrado vegetation of Brazil. *Botanical Review* 38:201–341.

Eiten, G. and Goodland, R. (1979) Ecology and management of semiarid ecosystems in Brazil. In: B.H. Walker (ed) *Management of semiarid ecosystems.* Elsevier, Amsterdam, pp. 277–300.

EMBRAPA/CPAC (1981) Relatório Técnico Anual do Centro de Pesquisa Agropecuária dos Cerrados (1979–1980). Planaltina DF Brazil.

Fry, C.H. (1983) Birds in savanna ecosystems. In: F. Bourlière (ed) *Ecosystems of the world: Tropical savannas.* Elsevier, Amsterdam, pp. 337–57.

Garrido, W.C., Silva, E.M., Junior, M.H.J. and Souza, O.C. (1979) Water-use efficiency by wheat varieties in a Cerrado soil of Brazil. In: R. Lal and D.G. Greenland (eds) *Soil physical properties and crop production in the tropics.* J.Wiley, New York, pp.215–25.

Goedert, W.J. (1983) Management of the Cerrado soils of Brazil: A review. *Journal of Soil Science* 34:405–28.

Goedert, W.J. (1990) Estratégias de manejo das savanas. In: G. Sarmiento (ed) *Las sabanas Americanas: aspectos de su biogeografía, ecología y utilizacíon.* Acta Científica Venezolana, Guanare, Venezuela, pp. 191–218.

Goedert, W.J. and Lobato, E. (1987) O solo como base dos sistemas de produção agrícola. *Anais do XXI Congresso Brasileiro de Ciência do solo.* Campinas, Brazil.

Goedert, W.J., Lobato, E. and Wagner, E. (1980) Potencial agrícola da região dos Cerrados brasileiros. *Pesquisa Agropecuária Brasileira* 15:1–17.

Goedert, W.J., Lobato, E. and Resende, M. (1982) Management of tropical soils and world food prospects. *Proceedings of 12th International Congress of Soil Science.* New Delhi, India, pp. 317–25.

Goodland, R. and Pollard, R. (1973) The Brazilian Cerrado vegetation: A fertility gradient. *Journal of Ecology* 61:219–24.

Gusmão, R.P. (1988) A expansão da agricultura e suas consequências no meio ambiente. In: S.T. Silva (ed) *Brasil uma visão geográfica dos anos 80.* Fundação Instituto Brasileiro de Geografia e Estatística, Rio de Janeiro, pp.323–32.

Hees, D.R., de Sá, M.E.P.C. and Aguiar, T.C. (1987) A evolução da agricultura na região Centro-Oeste na década de 70. *Revista Brasileira de Geografia* 49:197–257.

Heringer, E.P., Barroso, G.M., Rizzo, J.A. and Rizzini, C.T. (1977) A flora do Cerrado. In: M.G. Ferri (ed) *IV Simpósio sobre o Cerrado. Bases para utilização agropecuária.* São Paulo University Press, São Paulo, pp. 211–32.

IBGE (1960–1987) *Anuário estatístico do Brasil.* Fundação Instituto Brasileiro de Geografia e Estatística, Rio de Janeiro.

Lopes, A.S. and Cox, F.R. (1977) A survey of the fertility status of surface soils under Cerrado vegetation in Brazil. *Proceedings of Soil Science Society of America* 23:221–4.

281

Lucarelli, H.Z., Innocencio, N.R. and Fredrich, O.M.B.L. (1989) Impacto da construção de Brasilia na organização do espaço. *Revista Brasileira de Geografia* 51:99–138.

Mahar, D. (1989) Government policies and deforestation in Brazil's Amazon region. World Bank Publications, Washington DC.

Mesquita, O.V. (1989) Agricultura. In: A.C. Duarte (ed.) *Geografia do Brasil, região Centro-Oeste* (v.1). Fundação Instituto Brasileiro de Geografia e Estatística, Rio de Janeiro, pp. 149–70.

Morais, M.V.R. (1985) Processos erosivos nas encostas do Gama, Distrito Federal. *Revista Brasileira de Geografia* 47:417–26.

Mueller, C.C. and Penna, J.A. (1978) *Diagnóstico geo-sócio-economico da região Centro-Oeste do Brasil.* Brasilia University and Superintendência do Desenvolvimento do Centro-Oeste, Brasilia.

Ojasti, J. (1983) Ungulates and large rodents of South America. In: F. Bourlière (ed) *Ecosystems of the world: Tropical savannas.* Elsevier, Amsterdam, pp.427–439.

Ojasti, J. (1990) Las comunidades de mamiferos en sabanas neotropicales. In: G. Sarmiento (ed) *Las sabanas Americanas: Aspectos de su biogeografía, ecología y utilizacíon.* Acta Científica Venezolana, Guanare, pp. 259–93.

Parada, J.M. and Andrade, S.M. (1977) Cerrados: Recursos minerais. In: M.G. Ferri (ed) *IV Simpósio sobre o Cerrado. Bases para utilização agropecuária.* São Paulo University Press, São Paulo, pp.195–209.

Pastore, A.C. and Alves, E.R. (1984) Reforming the Brazilian agricultural research system. In: L. Yeganiantz (ed) *Brazilian agriculture and agricultural research.* EMBRAPA/DDT, Brasilia, pp. 117–28.

Ratter, J.A. (1971) Some notes on two types of Cerradão occurring in northeastern Mato Grosso. In: M.G. Ferri (ed) *III Simpósio sobre o Cerrado.* São Paulo University Press, São Paulo, pp. 100–2.

Ritchey, K.D., Souza, D.M.G., Lobato, E. and Souza, O.C. (1980) Calcium leaching to increase rooting depth in a Brazilian savanna oxisol. *Agronomy Journal* 32:40–4.

Sarmiento, G. (1983) The savannas of tropical America. In: F. Bourlière (ed) *Ecosystems of the world: Tropical savannas.* Elsevier, Amsterdam, pp.245–88.

Scolari, D.D.G. (1986) A evolução da produção agrícola na região dos Cerrados: Alguns índices. CPAC Documentos n. 23. EMBRAPA/CPAC: Planaltina DF Brazil.

Sick, H. (1965) A fauna do Cerrado. *Arquivos de Zoologia do Estado de São Paulo* 12:71–93.

Smith, N.J.H. (1982) *Rainforest corridors: The Transamazon colonization scheme.* University of California Press, Berkeley.

Soskin, A.B. (1988) *Non-traditional agriculture and economic development: The Brazilian soybean expansion, 1964–1982.* Praeger, New York.

Southgate, D. (1991) Policies contributing to agricultural colonization of Latin America's tropical forests. Unpublished report to the World Bank's Agriculture Department, January 1991.

Souza, M.A.A. (1990) Relação entre as atividades ocupacionais e a qualidade da água. In: M.N. Pinto (ed) *Cerrado: Caracterização, ocupação e perspectivas.* Brasilia University Press, Brasilia, pp.181–204.

Wagner, E. (1981) Política de ocupação e utilização dos Cerrados. CPAC Documentos n. 1. EMBRAPA/CPAC: Planaltina DF Brazil.

World Bank (1982) *A review of agricultural policies in Brazil.* World Bank Publications, Washington DC.

CHAPTER 14

TOWARDS THE DIVERSIFIED USE OF AUSTRALIA'S SAVANNAS

J.H. Holmes and J.J. Mott

SUMMARY

Europeans began grazing domestic cattle in Australia's savannas around 100 years ago. Displacing Aboriginal people, the initial goal was to "develop" the north which displaced Aboriginal people. Massive amounts of money were invested in rural development, most of which has been an abject failure. Considerable land degradation has occurred within Australia's savannas. Although technological advances have improved the delivery of basic services like health care and communications, they have not helped increase agricultural or pastoral output. It is now accepted that subsidising rural production in economically remote areas does not lead to self-sufficient development.

Recognition of the region's ecological carrying capacity and economic disadvantages has led to a new emphasis on conservation, alternative forms of land use (e.g. tourism) and social justice for Aboriginal people. New forms of land tenure and the reallocation of resource rights are now seen as the main vehicles for achieving more equitable, efficient and environmentally sustainable outcomes. However, the mixing of common-property and leasehold-ranching systems is creating some problems. The devolution of rights to local people to control land use has brought a new challenge: to ensure that local and broader community objectives do not come into conflict.

INTRODUCTION

Even by Australia's low levels of land-use intensity and population density, the Australian tropical savannas appear underutilised and underpopulated.

This inaccurate perception of "underdevelopedness" fails to recognise the ecological and economic constraints that characterise this region. Increasingly, however, these savannas are being recognised as places of limited agricultural potential. The persistent failure of past projects has brought about a shift in policy emphasis towards the pursuit of social justice for locationally disadvantaged peoples, restoration of Aboriginal rights, conservation, sustainable development and, most recently, deregulation and cost recovery on public services.

Regional definition

In this chapter we have restricted our coverage to the Australian northern and north-eastern savannas, north of the Tropic of Capricorn. We adopt a boundary generally slightly inland of Perry's (1968) boundary for economic cropping and sown pastures, that is often used to define the outer boundary of Australia's arid rangelands (see Figure 14.1).

In the ensuing discussion, we differentiate between the northern savanna zone, dominated by monsoon tallgrass savanna, and the north-eastern zone dominated by tropical tallgrass savanna. Their differences are attributable only partly to differences in pastoral potential and are more influenced by accessibility and regional infrastructure constraints.

In many reviews, the wooded monsoon tallgrass savannas of far northern Australia have been compared with the broad plains of the east African savannas. The low altitude northern Australian systems, however, appear much more infertile and support less primary production (Braithwaite 1990). In fact, climatically Australia's northern savannas are more similar to the Sahelian west African savannas (Williams and Probert 1983). Although infertile, these Australian systems show a high biodiversity of small mammals and birds and remain an important focus for conservation (Woinarski and Braithwaite 1990).

Comparison of Australian savannas with South American savannas suffers because the available information refers to the relatively tree-free savannas in Venezuela and neighbouring areas. Nevertheless, it would appear that the extremely high aluminium status of many Latin American soils maintains a savanna physiognomy at rainfalls much higher than similar structures in both Australia and Africa (chapter by Klink et al. in this book). The seasonal climate of the wooded Cerrado vegetation, however, has many similarities with Australian systems, even though these wooded savannas appear to have a higher potential for cattle production.

284

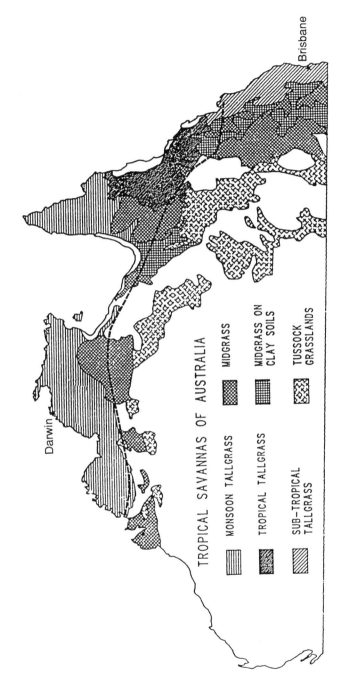

Figure 14.1 Main vegetation types in the Australian tropical savanna zone. The arid-humid line follows Perry (1967). Source: Mott, Williams, Andrew and Gillison (1985)

Indicators of laggard development

Both the northern and the north-eastern savanna regions of Australia have limited economic and population growth (Figure 14.2). With 15.4 per cent of Australia's area, they contain only 0.8 per cent of Australia's population. Their contribution to the national economy is equally modest. The region, however, does have a high proportion of Aboriginal people who, between 50 and 100 years ago, lived throughout the region as semi-nomadic hunters and gatherers reliant upon wildlife and native plants for their survival. They did not cultivate crops, nor keep domestic livestock.

Extending over an area of 877,180 sq km, the northern zone contained only 139,000 persons in 1986 at an average density of 0.16 persons per sq km (Table 14.1). The interior of the north-eastern savanna zone, embracing all areas over 50 km from the sea, is also sparsely settled, and contains only 54,600 persons at an average density of 0.18 per sq km. The adjacent narrow coastal zone, with a mix of tropical rainforest and savanna bioclimatic zones, is more intensively settled and contains 423,000 persons at an average density of 7.69 per sq km.

Over 25 per cent of the northern zone population is Aboriginal, which is the highest proportion in any Australian bioclimatic zone. Aboriginal people comprise over 80 per cent of the population in rural localities, and approximately 33 per cent of the dispersed rural population, but barely 10 per cent of the urban population. Around half (52 per cent) of the zone's population is located in Darwin. Of the remaining population, an unusually high proportion resides in rural localities with between 200 and 1000 persons. These are almost entirely Aboriginal communities. The dispersed rural population is very small and numbers only 26,000 persons. Of these, less than 3000 are directly employed in pastoral and agricultural production. The cattle industry has only modest labour needs, with over 1000 head of cattle for each equivalent full-time worker, including both permanent and casual labour.

Crude estimates of regional income generated by the various sectors indicate that, for the northern zone, mining generates five times, and tourism two times, more income than agriculture and pastoralism combined. Transfer payments from federal government sources to support regional infrastructure, social welfare and Aboriginal assistance far exceed in value the income gained from agriculture and pastoralism. The Northern Territory receives per capita grants three times higher than those in the most populous states, thereby enabling the public sector "broadly defined" to support 40 per cent of the Territory's workforce (Caldwell 1985).

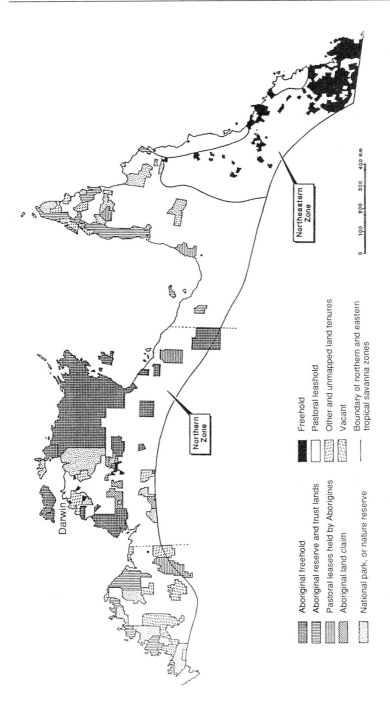

Figure 14.2 Land tenure in the Australian tropical savannas 1990. (Sources: Land tenure maps for Western Australia, Northern Territory and Queensland, with recent amendments from Lands Departments)

Table 14.1 Population of the Northern Savanna Zone. Source: Commonwealth of Australia Census

Place of residence	Aboriginal and Torres Strait Islanders		Non-Aboriginals		Not stated		Total	
	n	%	*n*	%	*n*	%	*n*	%
Urban	9972	29	80,957	81	3363	70	94,292	68
Rural locality	15,125	44	3531	4	325	7	18,981	14
Rural	9362	27	15,357	15	1146	24	25,865	19
Total	34,459	100	99,845	100	4834	100	139,138	100

Socio-economic and locational constraints

The imposition of a British-derived mode of occupancy in northern Australia has been fraught with difficulty. Most Europeans in the workforce failed to identify with "the North", regarding it as a remote, hostile, climatically unpleasant frontier zone, suited only to a "tour of duty" or as an opportunity to make money before going south. For many decades, transience retarded the rate of accumulation of needed experience to overcome northern problems, although this has changed markedly as northern living standards and lifestyle have improved. Unfamiliarity with a tropical environment and ignorance of appropriate agronomic and pastoral practices were compounded by an inability to subsume the Aboriginal economy and labour force into a colonial mode of exploitation. The only exception was the incorporation of many Aborigines into the cattle industry as stockmen and houseworkers.

While technological advances like refrigerators, radio communications, flying doctor services and education by correspondence have enhanced the quality of life in the north, the potential of technology to boost pastoral and agricultural production remains limited. Environmental difficulties continue to be reinforced by the economic and social burdens imposed by locational disadvantage. Without subsidies supported from taxation revenue collected elsewhere, agricultural output is dependent upon volatile, increasingly protectionist world markets, in which hinterland economies such as Australia's northern savannas are greatly disadvantaged. Intensification is economically unsound, given the extra cost burdens imposed by remoteness, additional to those already sustained in Australia's westernised, high-wage society.

ENVIRONMENTAL CONSTRAINTS

Soils

Although the soil pattern is complex in many areas, the range of soils present is not great (Williams et al. 1985). There is a high proportion of shallow-stony soils on moderate to steep slopes and considerable areas of sandy soils which have high content (60 per cent or more) of ironstone gravel throughout the profile. The most widespread soils with agricultural potential are the massive sesquioxidic soils (alfisols and oxisols). These include the red earths, yellow earths and grey earths and soils. In addition to these major groups, there are more limited areas of cracking clays. All have the long history of weathering, with the result that they are extremely low in nitrogen and phosphorus content and often have low water-holding capacity (Slatyer 1954; Isbell 1986). The exceptions are the cracking clays, especially those that formerly supported dense *Acacia* woodlands.

In addition to poor water-holding capacity, many of the earths show the development of surface crusts (Arndt 1965; Mott et al. 1979). Such crusting can markedly reduce infiltration and may provide a high strength barrier to seedling emergence. In addition to this physical restriction, the combination of high solar radiation fluxes and early wet-season storms can produce high (40–60°C) seed bed temperatures that restrict germination at a time when crop establishment is taking place (McCown et al. 1980).

Erosion

In a recent review of land degradation in Australia, over 92 per cent of land under cropping in the northern savanna regions was assessed as affected by water erosion (Clarke 1986). The high intensity summer rainfall that occurs in these savannas makes run-off and erosion a serious problem under conventional cultivation practice. In coastal regions losses of over 380 tonnes/ha per annum have been reported from cultivated lands.

Animal and pasture productivity

Domestic livestock were introduced into these savannas during the late 1880s and since that time there has been unrealistic optimism about their pastoral potential. Much of the philosophy has followed that of Francis Cadell (1866), who observed that "pastorally speaking, for the present it is a cattle country, feed most luxuriant, but too coarse for horses, too much grass seed for sheep, and subject to paroxysmal floods in the north-west monsoon".

While many socio-economic problems faced early pastoralists, it is recognised that a major cause of low carrying capacity and animal turn-off is the poor quality of forage during the dry season. Management problems in many areas of the monsoon tallgrasslands are reinforced by both wet-season and dry-season impediments to cattle control. Low livestock densities, combined with unfavourable terrain and plant cover, create serious problems in finding, mustering, moving and controlling cattle (Laut and Nanninga 1985). Pastoralists have responded by introducing *Bos indicus* blood into the cattle, trying to introduce nitrogen-fixing leguminous species into the pasture, and supplying feed supplements to cattle in an attempt to rectify nitrogen and phosphorus deficiencies (Siebert et al. 1976; Gillard and Winter 1984). These techniques worked well under experimental conditions (Tothill et al. 1985) and led to optimistic statements such as those by Begg (1972), who calculated that as much as 40 million ha of northern Australia could be sown to *Stylosanthes humilis*. Winter (1978) estimated that there was potential for a sixfold increase in animal production from improved animal husbandry and legume introduction. Implementation, however, was a different story. Major problems arose with pathogens (anthracnose on *Stylosanthes* spp.) and establishment costs were too high (McCosker and Emerson 1982; Mott 1987). Thus, although there has been a recent marked increase in the area sown to legumes, the dominant exotic species remain grasses. They exist either as volunteers, as in the case of *Bothriochloa pertusa* on degraded tropical tallgrass systems in the Burdekin catchment, or sown pastures such as *Cenchrus ciliaris* and *Panicum maximum* on the fertile clay soils formerly under the *Acacia* woodlands of the brigalow belt.

The philosophy behind improving savanna productivity has focussed on the animals, not on the surrounding savanna ecosystem (Mott 1987; Gardner et al. 1990). Herd numbers increased, with better-adapted animals surviving droughts, and these higher-producing animals also had much greater forage intake. Under these conditions and following a slump in cattle prices in the early 1970s, cattle herds rapidly increased in numbers, passing the "safe" carrying capacity of the native grasses in many regions (Figure 14.3). Major land and pasture degradation occurred (Williams 1991).

Land degradation

Pasture and land degradation, largely from overstocking in dry years, has been observed in many parts of the monsoon tallgrass savannas. Quantitative estimates of the area of degraded pasture land are difficult to obtain, but it is unlikely that this is less than the 13 per cent conservatively

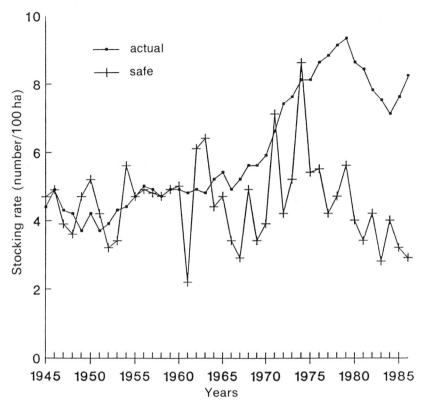

Figure 14.3 Actual stocking rate of beef cattle in the Dalrymple Shire of Queensland (tropical tallgrass savanna), compared with safe stocking rate. Source: Pressland and McKeon 1990

estimated by Wood and Fukai (1983) and it could be much higher. Williams (personal communication) estimates that more than 30 per cent of the Burdekin catchment is severely degraded, often losing more than 20 tonnes per ha per annum of soil from soil erosion. Similar broad-scale erosion has taken place in catchment areas serving the Ord River Dam in northern Western Australia.

Pasture stability

In addition to the degradation occurring under grazing use in the more infertile landscapes of the northern savannas, difficulties have been experienced in maintaining the stability of the introduced grass pastures on more fertile vertisols. Within 5–10 years of clearing of brigalow, aided by considerable taxation advantages, and planting of exotic grasses with high pasture yield, for example, there was a marked run-down in both pasture yield and quality and subsequent animal production (Henzell 1968; Catchpool 1984; Robbins et al. 1987). Recent research has shown that this apparent pasture run-down was due to the system returning to its normal equilibrium condition after disturbance (Myers and Robbins 1991). The reduction in pasture yield was due to a severe deficiency in available nitrogen because of limited mineralisation of humic material, immobilisation and slow turnover of nitrogen in plant residues with a high lignin : nitrogen ratio, and the gradual loss of nitrogen from animal excreta.

In the evaluation of pasture degradation, a critical factor is the ability of the pasture to regenerate after overgrazing. Studies in the *Themeda triandra*-dominated monsoon tallgrass savannas have shown that the soil sealing, which rapidly follows overgrazing, reduces the ability of the grasses to germinate from seed and that it can take a long period of destocking before effective regeneration takes place (Mott et al. 1979, 1985). In the tropical tallgrasslands there is a high risk of heavy rainfall immediately following drought periods.

In southern subtropical savannas, killing the tree upper storey has been found to increase grass growth and animal production (Tothill et al. 1985). Similar but smaller increases in forage production result from tree killing in the tropical savannas (Winter et al. 1990; Gillard et al. 1989). In deep permeable soils, however, water extraction is significantly reduced by tree killing, leading to increased run-off (Gillard et al. 1989) and changes in hydrology that lead to salinisation and waterlogging in the landscape (Williams and Coventry 1979). Policy makers are now beginning to recognise the need to control clearing so as to avoid salinisation problems similar to those resulting from the indiscriminant tree-clearing that has occurred in southern Australia.

Weed invasion

The invasion of inedible woody weeds into overgrazed land is increasing across much of the northern savannas (Tothill et al. 1982). Without either regular burning due to lack of fuel, or competition from perennial grasses, there has been a great increase in species such as Chinee apple (*Sisiphus mauritiana*) and rubber vine (*Cryptostegia grandiflora*). Both *Parkinsonia aculeata* and *Mimosa invisa* are forming impenetrable thickets in riverine areas and along flood plains (Dorney 1990). In clay soils, *Acacia nilotica* has infested over 25 per cent of arid tussock grasslands (Figure 14.1) and has the capacity to move into much of the midgrass savannas on more fertile clay soils (Carter 1990).

Conservation values

While the biomass of vertebrates is less than that in many of the commonly visited African park systems, the northern savannas remain a region of high biodiversity (Braithwaite and Werner 1987; Woinarski and Braithwaite 1990). As yet, there has been little loss of animal species from the northern monsoon tallgrass systems. Although there are recent dramatic increases in some weed species, great efforts are being made to control weed invasion in major park areas such as Kakadu (ANPWS 1986). The Cape York region still remains a large relatively untouched savanna system with, as yet, minimal disturbance from livestock production.

Such a reservoir of biodiversity is rare on any continent and could be of immense value in reclaiming degraded savannas in other parts of the world (Mentis and Tainton 1981). In other extensive grazing systems, like the sparsely populated Central Australian region, there has been a 60 per cent decrease in the number of species of small rodents and marsupials since pre-European times (Morton and Baynes 1985). Recognition of the importance of these animals to Aboriginal people, and for tourism and the maintenance of biodiversity values, is leading to a gradual change in development philosophy, but there is still a risk that the outcome could be the same as that which has occurred elsewhere.

LAND-USE CHANGE SINCE 1950

Since mid-century, there have been divergent land-use trends between the two zones. The north-eastern zone has experienced substantial gains in rural output. Rural land use continues to be dominated by pastoral activities, with modest expansion of cropping, primarily on clay soils.

The northern savanna zone is experiencing considerable change in land tenure and land use. Ongoing pastoral and agricultural failures have led to a rapid conversion of more marginal lands to Aboriginal ownership and national parks. It is likely that this trend will continue into the 1990s.

Land tenure policies

Traditionally, Australian governments have used land-tenure arrangements as a vehicle to encourage land development and rural population growth. The model, which began with closer settlement legislation in the 1860s, entails a progression through a series of leases involving higher rentals, more stringent development covenants, subdivision into smaller holdings, greater rights to compensation and longer lease periods culminating in the award of a freehold title.

Table 14.2 Land tenures in the Northern Savanna Zone 1990. Source: calculated from Figure 14.2

Tenure	Western Australia (km²)	Northern Territory (km²)	Queensland (km²)	Total (km²)	%
Aboriginal freehold	–	137,498	–	137,498	16
Aboriginal reserves*	23,790	–	16,418	40,208	5
Aboriginal claim	–	8540	–	8540	1
Aboriginal pastoral	5490	5195	4165	14,850	2
National parks etc.	19,520	29,428[†]	18,181	67,129	8
Pastoral lease	71,980	209,739	277,312	559,031	64
Vacant	23,180	–	–	23,180	3
All other	3050	13,664	10,030	26,744	3
Total	147,010	404,064	326,106	877,180	100

*Also includes Deeds of Grant in Trust (Queensland)

[†] This total includes 10,219 sq km of Aboriginal freehold land leased for National Parks purposes, and not shown in the area under Aboriginal freehold in this table.

These policies have been successful in achieving agricultural development in the more productive southern and eastern agricultural regions, including, to a certain degree, the north-eastern savanna zone. The northern savanna zone, however, has remained frozen at an early stage in this assumed progression and appears incapable of supporting unsubsidised forms of crop production (Table 14.2). Thus, the occupied pastoral lands of northern Australia are still held under low-rent, fixed-term leases, usually of 1000–4000 sq km, with very modest development and minimum-stocking covenants (often not fulfilled). By developed country standards, most northern savanna leases are barely able to yield sufficient income to support one family, together with casual labour and helicopter hire for a limited mustering effort. Recognition that extensive agricultural development is an elusive goal is leading to acceptance of the idea that the leasehold-tenure system may be adapted to serve newly acknowledged, land-management and land-use goals, such as Aboriginal land use, conservation, sustainable use, tourism and recreation (Young 1983).

In all states and the Northern Territory, pastoral leasehold tenures are being, or have recently been, subject to a detailed policy review. In all cases, the recommendations emphasise the need for sustainable management and the accommodation of non-pastoral activities. Less attention is being given to the need to require land development, capital investment and minimum stocking levels. New pastoral lease legislation, proposed for the Northern Territory, for example, will require pastoral leaseholders to maintain the "condition" of their pastures and place responsibility for land care on the landholder. The requirement to maintain a minimum number of domestic livestock will be repealed.

Dryland cropping

In addition to the soil characteristics constraining cropping already discussed, the absolute area suitable for high-technology cropping is much less than initially expected in northern Australia. Today, only 30 per cent of the soils originally estimated as being suitable for cropping are still grouped in that category.

Perhaps the most pertinent evidence of the failure of dryland cropping in the northern savannas is shown by the miniscule change in the actual area under cropping. Despite considerable investment in infrastructure development and research over the last 30 years, the area under crops in the Northern Territory has only increased from 1400 ha to 8000 ha. This is equivalent to some 5 per cent of the available land area considered technically suitable. In Queensland the difference is even more marked: only 0.5 per cent of land assessed as potentially viable for dryland crops is cropped.

Irrigated crops and pastures

While there have been several unsuccessful irrigation projects, by far the most ambitious proposal and conspicuous failure was the construction of a dam on the Ord River in the Kimberley District of Western Australia. Completed in 1972, the aim was to establish a large irrigation area but, as with other attempts to establish agriculture in the northern savannas, the excessively high cost burdens associated with farming, transport, storage and marketing, together with persistent difficulties with pests, have defeated efforts at growing cotton and other export crops. The large reservoir remains unutilised save as a recreation resource and tourist attraction. A handful of small farmers continue the unequal struggle against high costs and inadequate markets (Bauer 1964; Davidson 1966; Courtenay 1982; Mollah 1982). A multiplicity of reasons has been invoked to account for failure: inexperience, inadequate research, faulty planning, bad management, bad luck, undercapitalisation, overcapitalisation, inappropriate agronomic practices, inappropriate equipment, inadequate maintenance, lack of persist-

ence, lack of governmental support, lack of markets and high costs. But
the real reason is the same as the one that has retarded expansion of the
cattle industry, namely the high costs and low returns associated with a
remote, frontier region.

Aboriginal lands

In non-Aboriginal eyes, Aboriginal failure to exploit resources for commercial gain has been seen as "sufficient reason to deny their rights of prior
ownership, and to justify the alienation of all northern lands perceived to
be of economic worth. In the subsequent thrust to realise the potential of
these lands and their resources, Aborigines have played a relatively minor
role and their rights and priorities have been generally ignored" (Young
1988).

Pastoral occupancy provided the means for the dispossession of Aboriginal peoples, to a greater extent than occurred along the cold, northern
margins in Canada, Alaska and Scandinavia, where indigenous modes of
resource use, based on fishing and hunting or trapping fur-bearing animals,
were more readily subsumed into the commercial economy. In Australia,
with no comparable opportunity for subsumption, the Aboriginal economy
was destroyed and the Aboriginals dispossessed, even in areas where
pastoralism was based tenuously upon irregular "harvesting" of semi-feral
cattle.

Accordingly, the only major Aboriginal reserves were confined to tracts
of land unattractive to pastoralists, largely within the arid interior and the
northern savanna. In the northern savannas, major reserves were restricted
to the most rugged and remote lands in the northern Kimberley, Arnhem
Land, Cape York Peninsula and offshore islands (Figure 14.2). But over
the last two decades, Aboriginal land rights have become a matter of major
national concern, and, as indicated earlier, remarkably rapid changes have
begun to occur. Reserves have been transferred to Aboriginal group
ownership, national parks have been transferred to them then leased back
to the community, vacant crown land has been issued to them and pastoral
leases purchased for Aboriginal communities. A problem remains: while
Aboriginal communities have some interest in any economic benefits from
these forms of land use (of which mineral royalties offer the greatest
potential rewards), European experience has demonstrated that it is
unrealistic to assume that these lands can provide an adequate economic
base for their existing populations.

Thus, with the exception of those who have access to mineral royalties,
sources of cash income are very limited and the economic base of Aboriginal
communities will continue to be founded mainly on welfare payments,

together with public funding of housing and basic services. Further income for a few individuals is obtained from art and artefacts and, to a lesser extent, services to visiting tourists. Yet there remains a strong public expectation, shared by public officials in Aboriginal administration, that Aboriginal communities must become economically viable using Euro-centric criteria. But if White people cannot succeed in realising this elusive goal, it seems unrealistic to expect these people to do better.

Aboriginal people, however, see the issue differently. For them the prime value of the land is in helping to sustain critical elements in the Aboriginal culture, most notably: in subsistence foraging and hunting; in maintaining complex systems of inheritance, ceremonial and decision-making; and in custodianship and group identification. For these purposes the relatively unchanged nature of the country and its biota are of critical importance.

Emerging dualisms

As a result of the changes that failures to expand pastoral and agricultural land have facilitated, a new dualism has emerged between Aboriginal and non-Aboriginal occupancy. To Aborigines, land is not merely a "factor of production ... (but) ... a factor of existence ... (providing) ... religious significance, cultural integrity and social identification ..." as well as a resource base for traditional activities (Coombes et al. 1990). Land manage-ment and use on Aboriginal lands does not conform to Eurocentric goals. In the Northern Territory, but not Queensland nor Western Australia, Aborigines have been guaranteed the right to pursue their own land-use goals via an Act that issues a non-transferable freehold title to traditional landowners (see Box 14.1). These titles now cover 33 per cent of the Northern Territory and differ radically from the "land as commodity" or transferable-proprietary title that is granted to non-Aboriginal pastoralists. Communal land ownership, based upon traditional ties to the land, is leading to a restoration of traditional modes of land use and management and is creating new dualisms as Aboriginals follow new and different development paths not available to non-Aboriginal people. The dualism is increased by the access restrictions applied to non-Aboriginals and is requiring the development of new protocols on: mineral exploration and mining; the control of livestock diseases; feral animals and weeds; land conservation; and tourism.

This problem is particularly acute for the mining industry which, until recently, had always been given land-use priority, as was strikingly demonstrated by the transfer of large tracts of Aboriginal reserve into bauxite mining leases at Weipa in the 1950s.

Prior to granting the Northern Territory self-government in 1979, the federal government had been actively pursuing Aboriginal land rights within the territory, with programmes primarily based on the landmark Aboriginal Land Rights (NT) Act of 1976. After self-government the federal government retained responsibilities under this Act. The main elements are:

(1) Retention, for the time being, of federal powers over Aboriginal land claims, coupled with the creation of a new non-transferable Aboriginal freehold tenure held as a group title by Aboriginal Land Trusts;

(2) Rights for Aboriginal landowners to control access to their land and have "veto" over mining (subsequently modified);

(3) Rights for Aboriginal landowners to negotiate and then receive mineral royalties;

(4) Creation of Aboriginal Land Councils to administer Aboriginal lands in consultation with traditional owners;

(5) Transfer of existing Aboriginal reserves into the new non-transferable freehold title;

(6) Creation of the position of Aboriginal Land Commissioner to hear land claims to vacant crown land and to alienated land held by Aboriginal people, including pastoral leases; and

(7) Initiation of a programme of excisions of small areas from pastoral leases for use as living areas for Aboriginals currently residing on these leases.

The opportunity to make land claims (which will be terminated in 1997) has enabled Aboriginals to obtain further land outside their reserves, such that 34 per cent of the northern savanna lands in the Northern Territory is now under Aboriginal freehold title, with a further 2 per cent currently under formal claim. It seems probable that these claims will be granted.

Box 14.1 Aboriginal land rights

Tourism

The size and scenic diversity of the northern savanna zone are enough to guarantee some tourist activity in this area, but its isolation has, until recently, severely restricted tourism. Since about 1970, however, tourism has experienced very rapid growth and now ranks as a major sector in the regional economy.

There are considerable economic and social benefits in fostering tourism. It boosts the regional cash economy and increases employment opportunities. Moreover if managed wisely, tourism can be compatible with other forms of resource use, such as pastoralism, mining, conservation and traditional Aboriginal practices. Often tourism is perceived as reinforcing ecological arguments in support of declarations of national parks, nature reserves and wilderness areas. This smooths public acceptance of the retreat of pastoralism from submarginal lands. The economic reality, however, is different. Generous government guarantees for expensive, upmarket casinos and hotels in a few tourist enclaves have subsequently resulted in problems of overcapacity and ongoing deficits. Once again the burdens of long distances, inaccessibility, associated high travel burdens (in cost and time) and lack of adequate infrastructure have proved difficult to overcome. Furthermore, the rapid, unplanned growth of tourism, particularly at prime destinations, can lead to the destruction of those values that give the north a strong comparative advantage, namely its near-pristine scenery, its wildlife and the opportunity to be involved in an "outback" experience, selectively removed from some of the trappings of civilisation.

In less-developed regions, tourism can readily generate undesirable impacts, exceeding any potential economic benefits. Its marked seasonality and volatility devalues its worth as a basis for economic growth and job security. Furthermore, remote areas are severely disadvantaged by their lack of adequate infrastructure, which, in turn, invites two sharply divergent forms of tourism. At one extreme is the construction of *capital-intensive resort enclaves*, isolated from the local economy and society, with almost all inputs, including labour, being drawn from metropolitan sources. At the other extreme are the highly mobile, *self-sufficient safari-style groups*, mounted on vehicles ranging from motorcycles to four-wheel drives and safari tour-buses whose local purchases are mainly restricted to fuel and alcohol.

THE POLICY CONTEXT

The most important nationwide development programmes that have influenced savanna land use were introduced before the 1970s and include remote area income tax rebates, and substantial cross-subsidies for mail delivery, telephones, education, medical services, electricity and roads. Many of these cross-subsidies are now being phased out. Nevertheless, before 1970, a remarkably complex array of generous taxation concessions, especially on capital inputs and land development, subsidies on fertilisers and fuel, and low-interest loans, were introduced.

Zonally differentiated programmes

Recognising the heavy cost burdens imposed on remote producers, federal and state governments also adopted special assistance programs designed to reduce locational cost differentials. Of greatest importance in reducing production costs is the fuel-price equalisation scheme. This is of particular importance to pastoralists, especially in the northern savanna zone, where the subsidy is exceptionally high because of extreme isolation. On commercial pastoral enterprises, fuel costs account for 12.8 per cent of all variable and fixed costs (GRM International 1987) and also comprise the major component of off-farm transport costs, especially those associated with transporting cattle to distant markets and abattoirs.

In conjunction with the fuel-price equalisation scheme, problems of isolation and high transport costs have been attacked under a national Beef Roads Scheme that sought to construct single-lane, all-weather, bitumen roads, capable of providing prompt, flexible cattle transport by road-trains, throughout the northern savannas. At great national cost, these roads make it possible to quickly reduce cattle numbers and hence obtain the higher prices that cattle in better condition tend to attract. The benefits to tourists and other people who live in the northern savannas, however, may be greater than to those in the beef cattle industry.

Another concession is an income tax rebate that reduces the annual tax payable by people who live more than 250 km from a town of 2500 or more people by A$938. Other people within the northern savannas receive a rebate of $270. Given the smallness of the rebate and the avenues for income tax minimisation generally available to primary producers, however, it is unlikely that this allowance has had any measurable impact upon land use in the savanna zone.

In addition to the fuel-price equalisation subsidy, there are many other cross-subsidies imbedded in the delivery of basic rural services, such as mail delivery, telephone and electricity services that are supplied at a near-uniform charge, regardless of delivery cost. The element of cross-subsidy has escalated rapidly as new technologies have made service upgrading and expansion technically feasible, though at very high capital and operating costs. In each case the major benefit is improved quality of life for remote inhabitants rather than enhanced agricultural or pastoral production. Marked spin-offs in economic activity, employment and population growth have occurred, particularly in the towns within the sparselands but, in each case, this is paid for by the transfer payments from other parts of the economy (Holmes 1988a, 1988b).

APPROPRIATE GOALS AND RECOMMENDED POLICIES

Appropriate national and regional goals

A necessary, though not sufficient, prerequisite for effective policy formulation and programme development is the clear specification and ranking of major and minor goals. As indicated in Chapter 1, the overall goal of economic growth and development can only be pursued as far as it is consistent with:

(1) Efficient use of scarce resources;

(2) Ongoing ecological integrity of natural systems; and

(3) Socially equitable outcomes.

Given that northern Australia, in common with other tropical savanna zones, has limited "development potential", we consider that, whilst the drive for economic growth is important, it does not deserve a pre-eminent place in the formulation of policies that influence the use of Australia's savannas. Importantly, this region:

(1) Does not experience either the population pressures, or the low per capita incomes, that prevail in almost all other savanna zones;

(2) Does not have a comparative advantage in the pursuit of broad-acres cropping and pastoral development; and

(3) Contains many relatively unaltered ecosystems that are of national ecological significance and provide a significant opportunity to improve the position of Aboriginal people within Australia.

301

Emerging forms of savanna land use

Recognising these considerations and that the emerging dualism is likely to create sharply differentiated forms of land ownership, use and management, especially in the northern savannas, we can expect

(1) *Localised nodes of agricultural, pastoral and urban development*, involving further intensification of land use, based on investments in production systems and infrastructure that centre on places like the lower Ord Valley and the immediate Darwin hinterland;

(2) *Scattered development enclaves* based upon localised mining and tourist resources and exotic ventures such as spaceports;

(3) *Aboriginal homelands* centred on major tracts of Aboriginal land, held mainly under freehold title, whose prime function is to provide sustainable modes of land use in accordance with the wishes of landowners, presumably with large tracts utilised primarily for traditional activities;

(4) Manageable, but low-yielding *pastoral lands* capable of supporting a modest level of investment in fencing and waters to ensure basic herd control; and

(5) Extensive tracts of *"undeveloped" submarginal land*, posing ongoing challenges in land tenure, ownership, management and use.

Pivotal role of policies on land allocation and land-use planning

Recognising that the above land-use scenario is likely, by far the most important policy area relates to direct decisions about the reallocation of rights to use land, rights to exclude or control others who wish to use it, and all the obligations that attach to those rights. This direct focus on land "allocation" and land-use planning is consistent with current policy emphases in comparable marginal lands in Alaska (Morehouse 1984) and Canada (Freeman 1981; Department of Indian Affairs and Northern Development 1986; Fenge and Rees 1987). It is also consistent with the status of virtually all the other savanna regions described in this book. In all cases there are considerable political pressures, plus sound environmental, social and economic reasons, to equitably redistribute resource rights and modify the conditions and obligations that attach to those rights.

Already in north Australia, there have been emerging signs of a growing willingness of governments to participate directly in reshaping land allocation and land-use rights. On lands where a Eurocentric focus still prevails, all three northern administrations have shown some indication, albeit hesitant, of becoming more interventionist. For example:

(1) In Western Australia, the state government purchased eight major pastoral leases, comprising most of the prime grazing lands of the western Kimberley (mainly south of the savanna zone shown in Figure 14.2), and has begun to reallocate them amongst pastoral and Aboriginal interests.

(2) In the Northern Territory, a 1985 inquiry into land-use options for the Gulf District (Holmes 1986, 1990) has been followed up by the preparation of a draft land-use strategy involving the reallocation of pastoral land to other land uses and the development of new multipurpose leasehold tenures (Northern Territory 1990).

(3) In Queensland, there has been agreement between the state and federal governments to undertake a 5-year project in strategic land-use planning for Cape York Peninsula, in recognition of the uncertainties surrounding future land-use directions and the significance of the region for conservation and Aboriginal uses.

This political willingness to consider considerable change is possible for three reasons. First, a high proportion of land is either state-owned, or held under Aboriginal group title rather than under private title. Second, almost all of the privately-held land is under term pastoral leasehold tenure by which the state, at least nominally, still retains ownership, and does retain greater powers of intervention than on freehold land. Third, there is strong public interest in ensuring that the vast ecological significance of savanna lands of marginal pastoral potential are not degraded. In short, the individual private interests of the leaseholders weigh low in comparison to the "public interest".

Major development nodes

For the major development nodes, it will be critical to begin with a realistic appraisal of development prospects, and learn from the economically, socially and ecologically wasteful efforts that have characterised past attempts. The two main nodes are Darwin and its immediate hinterland extending south to Katherine and, secondly, the lower Ord Valley which still awaits the advent of large-scale, export-oriented, irrigation-based agriculture. There is an urgent need to manage the process of land conversion from low-value pastoral leasehold to higher-value uses, including hobby farming, rural residential, market-oriented agriculture, broad-acres recreation, and other urban-focussed uses. The task is made more difficult because initial, low land values encourage speculative and non-speculative expansion far beyond foreseeable needs. Strategies that guide the amount and location of land release into smaller holdings are essential. There is also a need to plan to minimise the potentially disastrous overspill impacts over an extensive hinterland beyond the development zone. Local population concentrations of highly mobile (with boats and four-wheel-drive vehicles), recreation-oriented, urban Australians who treat the wilderness as a free-for-all playground can have severe impacts on pastoralism, Aboriginal populations and significant and sensitive riverine, estuarine and other wetland habitats. Even stronger pressures are felt in eastern savannas adjoining the major east-coast population concentrations.

Developed enclaves

Currently the main developed enclaves are: Weipa (population 3135), a bauxite mine and port; Nhulunbuy (3514), also bauxite, including first-stage processing; Alyangula (1244), manganese mining; Jabiru (1408), uranium mining and tourism; Kununurra (3135), limited agriculture and tourism pivoting on the Ord Dam; and Argyle (under 1000), diamond mining. Further development enclaves are highly prospective, involving additional mining, stand-alone tourist resorts and the proposed spaceport on northern Cape York Peninsula.

Development enclaves occur fortuitously, usually in remote locations, focussing on a highly localised resource. The first three listed above are encircled by Aboriginal lands, while Jabiru is within Kakadu National Park and leased from Aboriginal owners. These enclaves pose very severe challenges, as yet unresolved: how to maximise local economic and social benefits, and minimise adverse social and environmental impacts.

One strategy, increasingly favoured, is to effect a near-complete divorce between the enclave and its encircling region, most successfully achieved by having a fly-in-fly-out workforce. Most of the workforce at the Argyle diamond mine, for example, work for 14 continuous days and are then flown to their homes in Perth for seven days' recreation. Perth is over 1100 km distant, but this arrangement makes Argyle a satellite of the distant metropolis.

In other ventures some sporadic efforts have been made to find jobs for local underemployed Aboriginals with very limited success. Where new mining proposals occur in Aboriginal lands, currently the main negotiation emphasis is on the size of royalty payments to Aboriginal people and on control of environmental impacts. Royalties can provide substantial income, but employment opportunities are available to only a few favoured Aboriginal people.

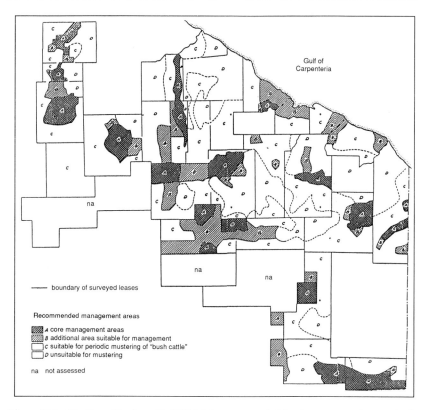

Gulf of
Carpenteria

na

na   na

—— boundary of surveyed leases

Recommended management areas

▨ *A* core management areas
▨ *B* additional area suitable for management
☐ *C* suitable for periodic mustering of "bush cattle"
☐ *D* unsuitable for mustering

na   not assessed

Figure 14.4 Northern Territory Gulf District: recommended herd management areas, bush mustering areas and unmusterable areas. Source: Holmes (1990)

Manageable pastoral lands

A delineation of manageable pastoral lands has yet to be undertaken across Australia's north, and the only purposeful attempt has been in the Northern Territory Gulf Investigation (Figure 14.4). Management areas were determined using five criteria: carrying capacity, musterability, "fenceability", internal accessibility and existing capital improvements. The result broadly coincides with areas carrying more than 1.56 cattle per sq km (Figure 14.5). These criteria, and all other relevant information, suggest that almost the entire north-eastern savanna zone comprises manageable cattle land, whereas, in the northern zone, less than half of the land currently held under pastoral lease can be classed as manageable.

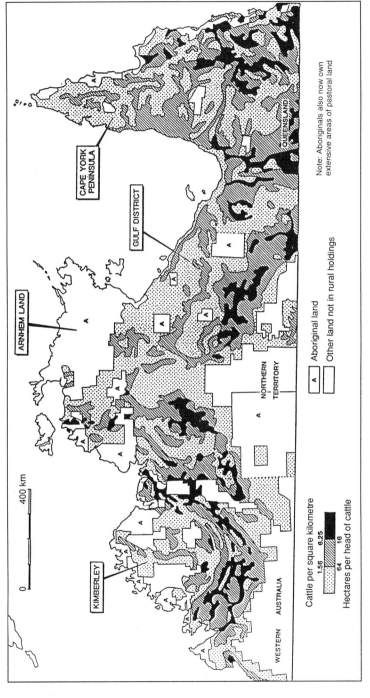

Figure 14.5 Density of beef cattle in north Australia. Source: Atlas of Australian Resources

Non-Aboriginal submarginal lands

Pastoral enterprises on submarginal land are dependent upon asset fixity, zero opportunity costs, the regenerative capabilities of the semi-feral cattle herd and the willingness of the leaseholder to pursue a simple lifestyle. Non-speculative land values are at or close to zero dollars per sq km. Usually, half-hearted, sporadic efforts at improvement of herds, fencing, water and pastures have been to no avail. Viability has been threatened by the national campaign to eradicate brucellosis and tuberculosis, which requires strict herd control via uneconomic investment in fencing, yards and waters. Given this reality, coupled with the large national gains from the complete eradication of brucellosis and tuberculosis, the most rational alternative is to totally destock these areas and examine other land-use options, including Aboriginal use, tourism, recreation and conservation. Whatever decisions are made, however, there will still be a need for sound land management in face of alien invasions of humans, plants and animals which threaten the prime value of these "unproductive" lands, namely their near-pristine condition. Thus major policy directions for these submarginal pastoral areas will need to involve considerable strategic land-use planning and the development of new forms of tenure and administrative techniques.

Aboriginal homelands

Since the mid 1970s, the map of Australian land ownership, and rights to use and deal with that land, have been dramatically transformed by the award of common-land titles to traditional Aboriginal owners in the Northern Territory and South Australia. Progress is anticipated in Western Australia and Queensland, spurred by the 1992 landmark decision of the High Court recognising Aboriginal traditional title over property, save where extinguished by the award of a grant or title to another party.

The two fundamental policy areas to be addressed are, first, the principles and mechanisms for determining Aboriginal land claims and, second, the scope for policies that promote sustainable land use on Aboriginal lands.

Satisfying Aboriginal land claims

Although elaborate procedures have been developed to make determinations on Aboriginal land claims, very serious deficiencies remain. Apart from the obvious inconsistencies among different states and the federal government, current policy on land rights can be criticised on two related

307

counts. First, the award of lands is constrained by the historical accident of land availability, either as Aboriginal reserve or vacant public land or national park, rather than through any informed appraisal of the balance between Aboriginal and other interests in the land. As a result, the outcome is often very inequitable. Second, the title being issued is simultaneously more powerful and more restrictive than those available to non-Aboriginal people who also live in the region. This reinforces the dualism between Aboriginal and non-Aboriginal lands and the associated social divisions are strongly felt locally; and it also strengthens the growing resistance towards recognition of further Aboriginal land claims, such that the lack of any compromise land title may exclude some Aboriginal groups from acquiring any land. Accordingly, it may reinforce and perpetuate inequitable outcomes for Aborigines, and preclude multiple and/or joint land-use options that require shared decision-making between Aboriginal and non-Aboriginal representatives.

Controlling land use

At a different level, land-use and land-management issues on Aboriginal lands have to be addressed. For policy makers and advisers there are fundamental conflicts between: the highly desirable but seemingly irreconcilable goals of sustainable land use; the elimination of welfare-dependency; and Aboriginal autonomy in resource management. As a starting point in resolving these dilemmas, the elimination of welfare-dependency should be recognised as unrealistic and granted low priority, certainly a much lower priority than recently awarded to it by the government. The resource base is already inadequate for existing Aboriginal populations, and will become increasingly so as populations grow and struggle with traditional decision-making procedures that do not conform with commercial practice. The management of cattle herds, for example, is influenced by traditional sustenance and ritual activities, and by perceptions of the appropriate balance between cattle and other traditional food resources. New technologies like four-wheel-drive vehicles, outboard motors and rifles have to be accommodated. It requires a supreme act of faith to assume that contemporary Aboriginal communities will engage in environmentally-sound custodial resource use, any more than non-Aboriginal land users have done in the past. Thus Aboriginal, like non-Aboriginal, land-use practices may need to be constrained. National land-use goals, for example the eradication of animal diseases and control of feral animals, may need to be pursued, irrespective of the desires of Aboriginal people.

Policies for Aboriginal participation outside homelands

Land-related policies for Aboriginal participation will need to be focussed on manageable grazing lands and undeveloped lands. Generally, policies in developed cores and enclaves will not be land-based, save in providing for Aboriginal control of community residential land and sacred sites.

Where traditional links to land have been maintained, there are opportunities for a more influential and rewarding involvement in a wide array of land-based activities, with primary income-earning activities being directed towards land custodianship and tourism. In these cases, Aboriginal people have an identifiable role in decision-making processes that can be accommodated. The arrangement whereby Kakadu and Uluru National Parks have been granted to their traditional owners, then leased back from them, has markedly enhanced the appeal of these parks to tourists, to the point where environmental tourism is becoming ancillary to Aboriginal cultural tourism, rather than the reverse. Comparable opportunities have yet to be fully realised elsewhere in Northern Australia.

Quite apart from the very positive outcomes for tourism through active Aboriginal involvement, one further reason for pursuing this policy is that poorly controlled, insensitive tourism can be more damaging environmentally and to Aboriginal society than any other economic activity.

It would involve more people engaged in activities difficult, if not impossible, to control, competing for wildlife and other resources still vital to Aboriginal livelihood, threatening sacred and other sites of great significance and in many ways providing deliberately or unconsciously the material for Black–White antagonisms (Coombes 1984).

The current challenge in fostering a revival of Aboriginal skills and motivation demonstrates the importance of integrating policies on land ownership, land use and land management with parallel policies on economic and social development of local populations, particularly Aboriginal groups who form the majority of the non-urban population in the northern savanna.

OTHER RECOMMENDED POLICIES

Given the lowly contribution of the Australian savannas to the national economy, it is inappropriate and unrealistic to shape national policies to serve the specific needs of tropical savanna producers. Nevertheless, it is important to identify the likely impact of national policies on savanna land use.

Trade policies

Currently Australia has adopted a leading international role as an advocate of free trade. Consistent with this advocacy, Australia has markedly reduced its tariff barriers on imports, particularly of manufactured goods and has been dismantling its modest package of agricultural assistance programmes. Primary producers still remain severely disadvantaged, with an ever-tightening cost-price squeeze, only partially remedied by more competitively priced farm inputs. Further trade liberalisation is likely to be to the net benefit of people living within Australia's northern savannas.

If the current round of General Agreement on Tariffs and Trade (GATT) negotiations is unsuccessful, as appears highly likely, there will be growing pressure on Australia to join a major trade block, of which the most likely is a prospective south-east and east Asian block. While there would be marked trade advantages for Australia within such a block, there remain doubts about its possible emergence. For Australian tropical savanna regions, this trading block would promise very significant benefits. The zone has a comparative advantage in broad-acre pastoralism and excellent prospects for marketing niches in specialised agricultural crops. There would also be benefits in obtaining lower-priced imports.

Macro-economic policies

The policy context is broadly the same in this area. In Australia macro-economic policies are increasingly shaped by world trade and international finance, with less attention being given to the needs of specific regions or sectors.

As stated earlier, income taxation policies may have had some influence on development in the tropical savannas, but this influence is not readily determined. The special zonal tax rebates are small, and of little consequence to primary producers. A general shift in the tax towards the taxation of value-added goods and/or resource use could disadvantage this region, particularly if accompanied by a reduction in the considerable freight and service subsidies that it already receives.

Agricultural policies

Within Australia, beef marketing has been, and remains, relatively unsubsidised and uncontrolled. Government intervention has been limited to essential tasks, such as inspection to satisfy domestic and export standards, together with bilateral negotiation for beef quotas in main export markets.

In the current deregulatory context, no group is arguing for more structured marketing arrangements, and any such policy proposal would be unrealistic and unwarranted. Government assistance towards improved product standards and the opening of new overseas markets seems to be the limit of intervention.

One possible policy goal would be to reduce uncertainty and variability in beef markets and prices. The northern beef industry has a very low and flexible cost structure, is highly resilient and capable of adapting to sharp price fluctuations, mainly by adjusting the ratio between cattle turn-off and herd build-up. This strategy, however, is not linked to seasonal conditions and has led to serious land degradation, most clearly revealed in the prolonged beef recession of the early and mid 1970s when a persistent increase in herd numbers initiated severe pasture degradation in many areas. Policies to remedy this range from providing incentives to minimise the expansion of livestock numbers during recessions, to the strict enforcement of leasehold covenants which aim to maintain vegetation and soil structure.

A case can be made for more actively supporting research, development and marketing programmes, but the progress that has been made so far has been disappointing. Past experience would suggest that the benefits from this form of public activity would not cover the costs. If a new Asian trading block emerges or there is a general move to globally freer trading arrangements, however, this situation could change.

Policies on tourism and conservation

Government policies in support of tourism, recreation, sustainable land use and preservation of natural habitats and wildlife are closely tied to land-allocation policies and Aboriginal-focussed policies. The relatively high biodiversity of Australia's northern savanna regions makes them both nationally and internationally important in the pursuit of conservation objectives.

Policies and programmes supportive of tourism need to be closely integrated with programs of pastoral land use, land conservation, Aboriginal land rights and social advancement, employment and training, and infrastructure provision. Local businesses and people, Aboriginal and non-Aboriginal, require training programmes and other forms of assistance to enhance local capacity to foster tourism on a mutually rewarding basis.

Rural and remote assistance policies

Special subsidies and pricing policies for the delivery of public goods and service, coupled with persistent underpricing of road charges for heavy transport vehicles, have helped people who reside within, or have invested in, Australia's savannas. Along with other Western countries, Australia is currently engaged in a series of major national policy reviews over deregulation, privatisation and full cost-recovery for basic services. For services such as post and telecommunications, a probable outcome is likely to be competition in the profitable intra-urban and inter-urban markets, with a continuing public monopoly in the unprofitable rural services, and with losses covered by direct subsidies rather than intrasystem cross-subsidies. As pointed out by then Prime Minister Hawke, the pursuit of cost-recovery would be "... a chase after a vanishing target. What would vanish would be most of the phones and mail services outside Australia's major cities" (*The Australian* 1986). Such an outcome is clearly contrary to entrenched national goals that require all people to have equal opportunities of access to basic social services.

The use of direct subsidy mechanisms may have desirable outcomes in leading to a closer scrutiny of cost-burdens and the introduction of disincentives for people to live in socially "expensive" locations. This can be achieved by making consumers meet at least part of the extra cost whilst providing rebates for deserving cases. There are similar policy problems with any proposal to abolish fuel transport subsidies. The removal of fuel-price equalisation across Australia would have a dramatic impact on remote peoples whose standard of living is heavily reliant on the maintenance of transport costs within reasonable bounds.

Social welfare policies

Again, a cornerstone of Australian social philosophy is the belief that society must provide a cushion against unemployment and give all unemployed adults, irrespective of race, an opportunity to receive a fortnightly welfare payment. Since the extension of this programme to include all Aborigines living within their communities in 1967, there has been a sharp increase in welfare-dependency. As a result, serious social problems – alcohol, violence, ill health and high suicide rates – have emerged. Solutions to these social ills are not readily proposed. There have been some promising community initiatives, with the two most effective being prohibitions on alcohol (with heavy penalties, not readily enforceable) and the redirection of unemployment benefits to the community council, with individual entitlements being paid on the basis of part-time community service work. One promising policy direction is to link unemployment benefits to an expanding array of training and part-time employment opportunities, with an expansion beyond unskilled service jobs into land and resource-based work using traditional Aboriginal skills. Work would be mainly in the areas of land custodianship and tourism.

RELEVANCE TO GLOBAL SAVANNA MANAGEMENT

We have recognised two basic points of similarity between northern Australia and other tropical savannas, namely bioclimatic affinities and low development potential. We have also emphasised the sharply divergent socio-economic context, with Australia's northern savannas being uniquely located within a Western, affluent nation. The relevance of Australian experience must be assessed within this context of basic similarities and differences. There is much common ground in biophysical research, especially in climate–soil–plant–animal relations and in management applications.

All savanna ecosystems can be considered to be non-equilibrial, dominated by stochastic patterns of drought (Ellis and Swift 1988; Solbrig, this book), from which man tries to extract a relatively constant yield. Under such conditions development strategies that confine pastoralists to grazing blocks or ranches reduce the spatial scale for exploitation of natural variability over large areas and can result in major land degradation in serious droughts.

Given the extent and nature of the land degradation that has occurred on Australian savanna lands, an initial reaction might be to follow the lead of Downing (1990) who stated that "apparent benefits bestowed on rangeland condition by transhumance in Africa suggest that traditional (European) pastoral management practices be opened for questioning". Opportunities for more flexible market-based grazing regimes require the development of production systems that remain within the ecological capacity of each savanna. Possibilities include improvements in mustering, transport and marketing. One option is to use the savannas as opportunistic sources of beef cattle to be fattened elsewhere.

Yet the parallel opportunities for more resilient management have yet to be realised. Indeed, the growing emphasis on on-farm feed supplementation is having highly adverse effects in promoting further inflexibility in stocking rates. In short, it is not clear that the market-based ranching system found in Australia is naturally superior in either an ecological or social sense to that found in other parts of the world. Those differences that do exist come from the capacity of the Australian economy to provide large transfer payments from other parts of its economy. Such a strategy is unlikely to be politically or economically sustainable in most of the other savannas described in this book.

Australia's pursuit of national economic growth can largely be achieved without heavy reliance on its savanna region. Indeed, overall growth rates may be improved by minimising governmental intervention. Experience has demonstrated that the savanna regions should not be awarded high priority in the pursuit of national development goals, and it is inappropriate for taxation, fiscal, marketing or other macroeconomic policies to be tailored to meet their particular requirements. This does, however, mean that local land administration, tenure and land-management policies must be used to counteract the adverse effects of national policies on local land-use practices.

Australia is exploring new and innovative land-tenure arrangements in an attempt to solve the social, economic and environmental problems that characterise all savannas. The attempt to use subsidies to raise productivity and make the region self-sufficient having failed, the emphasis has shifted to questions of social justice and equity. A dualistic approach has emerged which raises another set of important questions about how one mixes private and common-property systems. Perhaps it is ironic that, even in highly Westernised countries like Australia, there is a growing recognition that common-property systems have many desirable cultural and economic characteristics that are lost in private-property systems.

Finally, the recognition of Aboriginal aspirations, focussing on land rights and social self-sufficiency, may provide experience of relevance to some other nations with a comparable clash between indigenous minorities and an economically and politically dominant majority culture.

REFERENCES

Australian National Parks and Wildlife Service (1986) *Kakadu national park plan of management.* Australian Government Publishing Service, Canberra.

Arndt, W. (1965) The nature of mechanical impedence of seedlings by soil surface seals. *Australian Journal of Soil Research* 3:45–54.

Bauer, F.H. (1964) *Historical geography of white settlement in part of northern Australia.* CSIRO, Canberra.

Begg, J.C. (1972) Probable distribution of Townsville stylo as a naturalised legume in tropical Australia. *Journal of the Australian Institute of Agricultural Science* 38:158–62.

Braithwaite, R.W. (1990) Australia's unique biota: Implications for ecological processes. *Journal of Biogeography* 17:347–54

Braithwaite, R.W. and Werner, P.W. (1987) The biological value of Kakadu National Park. Search 18:296–301.

Cadell, F. (1866) *South Australian Parliamentary Papers.* 24:168–9.

Caldwell, P.J. (1985) Public sector in the Northern Territory economy. In: P. Loveday and D. Wade-Marshall (eds) *Economy and people in the north.* North Australia Research Unit, Darwin.

Carter, J. (1990) The ecology and control of *Acacia nilotica*. In: T. Armstrong (ed) *Noxious weeds workshop*. The Weed Society of Queensland, Brisbane, 58–62.

Catchpool, V.R. (1984) Cultivating the surface soil to renovate a green panic (*Panicum maximum*) pasture on a brigalow soil in south east Queensland. *Tropical Grasslands* 18: 96–9.

Clarke, A.L. (1986) The impact of Australian practices on Australian soils: Cultivation. In: J.S. Russell and R.F. Isbell (eds) *Australian soils: The human impact*. The University of Queensland Press, Brisbane, pp. 273–303.

Coombes, H.C., Dargavel, J., Kesteven, J., Ross, H., Smith, D.I. and Young, E. (1990) *The promise of the land: Sustainable use by Aboriginal communities*. Centre for Resource and Environmental Studies, Australian National University, Canberra.

Coombes, H.C. (1984) Tourists on black lands: Opportunity or threat? *The Age*, 20 December 1984, Melbourne.

Courtenay, P.P. (1982) *Northern Australia: Patterns and problems of tropical development in an advanced country*. Longman Cheshire, Melbourne.

Davidson, B.R. (1966) *The northern myth*. Melbourne University Press, Melbourne.

Department of Indian Affairs and Northern Development (1986) *Comprehensive land claims policy*. Minister for Supply and Services, Ottawa.

Dorney, W. (1990) Ecology and control of woody legumes. In: T. Armstrong (ed) *Noxious weeds workshop*. The Weed Society of Queensland, Brisbane, pp. 63–7.

Downing, B. (1990) Traditional pastoral systems compared: Transhumance in Africa against static grazing in Australia. *Range Management Newsletter* 90/3:16–8.

Ellis, J.E. and Swift, D.M. (1988) Stability of African pastoral ecosystems: Alternate paradigms and implications for development. *Journal of Range Management* 41:450–9.

Fenge, T. and Rees, W.E. (eds) (1987) *Hinterland or homeland? Land use planning in Northern Canada*. Canadian Arctic Resources Committee, Ottawa.

Freeman, M.M.R. (ed) (1981) *Proceedings of the First International Symposium on Renewable Resources and the Economy of the North*. Canada Man and Biosphere Program, Banff.

Gardner, C.J., McIvor, J.G. and Williams, J. (1990) Dry tropical rangelands: Solving one problem and creating another. In: D.A. Saunders, A.J.M. Hopkins and R.A. How (eds) *Australian ecosystems: 200 Years of utilisation, degradation and reconstruction. Proceedings of Ecological Society of Australia* 16:279–86.

Gillard, P. and Winter, W.H. (1984) Animal production from stylosanthes based pastures in Australia. In: H.M. Stace and L.A. Edye (eds) *The biology and agronomy of stylosanthes*. Academic Press, London, pp. 408–30.

Gillard, P., Williams, J. and Moneypenny, R. (1989) Clearing trees from Australia's semi-arid tropics – production, economic and long term hydrological changes. *Agricultural Science* 2:34–9.

GRM International (1987) *Northern Territory Pastoral Industry Study*. Northern Territory Development Corporation, Darwin.

Henzell, E.F. (1968) Sources of nitrogen for Queensland pastures. *Tropical Grasslands* 2:1–17.

Holmes, J.H. (1986) *The pastoral lands of the Northern Territory Gulf District: Resource appraisal and land use options*. Report to Northern Territory Department of Lands and Housing, Uniquest, Brisbane.

Holmes, J.H. (1988a) New challenges within sparselands: The Australian experience. In: R.L. Heathcote and J.A. Mabbutt (eds) *Land, water and people: Geographical essays in Australian resource management*. Allen and Unwin, Sydney.

Holmes, J.H. (1988b) Private disinvestment and public investment in Australia's pastoral zone: Policy issues. *Geoforum* 19:307–22.

Holmes, J.H. (1990) Ricardo revisited: Submarginal land and non-viable cattle enterprises in the Northern Territory Gulf District. *Journal of Rural Studies* 6:45–65.

Isbell, R.F. (1986) The tropical and sub-tropical north and northeast. In: J.F. Russell and R.F. Isbell (eds). *Australian soils: The human impact*. The University of Queensland Press, Brisbane, pp. 3–35.

Laut, P. and Nanninga, P.M. (1985) Landscape data for cattle disease eradication in northern Australia. Division of Water and Land Resources Technical Paper No. 47, CSIRO, Australia.

McCosker, T.H. and Emerson, C.A. (1982) Failure of legume pastures to improve animal production in the monsoonal dry tropics of Australia: A management view. *Animal Production in Australia* 14:337–40.

McCown, R.L., Jones, R.K. and Peake, D.C.I. (1980) Short term benefits of zero-tilling in tropical grain production. *Australian Agronomy Conference Proceedings*, Lawes, p. 220.

Mentis, M.T. and Tainton, N.M. (1981) Stability, resilience and animal production in continuously grazed sour grassveld. *Proceedings of Grassland Society of Southern Africa* 16:37–43.

Mollah, W.S. (1982) *Humpty Doo rice in the Northern Territory.* North Australia Research Unit, Darwin.

Morehouse, T.A. (ed) (1984) *Alaska resources development. Issues of the 1980s.* Westview Press, Boulder.

Morton, S.R. and Baynes, A. (1985) Small mammal assemblages in arid Australia: A reappraisal. *Australian Mammalogy* 8:159–69.

Mott, J.J. (1987) Planned invasions of Australian tropical savannas. In: R.H. Groves and J.J. Bourdon (eds) *Ecology of biological invasions: An Australian perspective.* Australian Academy of Science, Canberra, pp. 97–105.

Mott, J.J., Bridge, B.J. and Arndt, W. (1979) Soil seals in tropical tallgrass pastures of northern Australia. *Australian Journal of Soil Research* 30:483–94.

Mott, J.J., Williams, J., Andrew, M.H. and Gillison, A. (1985) Australian savanna ecosystems. In: J.C. Tothill and J.J. Mott (eds) *Ecology and management of the world's savannas.* Australian Academy of Science, Canberra, pp. 56–82.

Myers, R.J.K. and Robbins, G.B. (1991) Maintaining productive sown grass pastures. *Tropical Grasslands* 25:103–11.

Northern Territory Department of Lands and Housing (1990) *Draft Gulf region land use and development study.* Darwin.

Perry, R.A. (1967) The need for rangelands research in Australia. *Proceedings of the Ecological Society of Australia* 2:1–14.

Perry, R.A. (1968) Australia's arid rangelands. *Annals of the Arid Zone* 7:43–9.

Pressland, A.J. and McKeon, G.K. (1990) Monitoring animal numbers and pasture condition for drought administration: an approach. In *Working Papers: Range Monitoring Workshop. Proceedings of the Fifth Australian Soil Conservation Conference, Perth.*

Robbins, G.B., Bushell, J.J. and Butler, K.L. (1987) Decline in plant and animal production from ageing pastures of green panic (*Panicum maximum* var. *trichoglume*). *Journal of Agricultural Science* 108:407–17.

Siebert, B.D., Playne, M.J. and Edye, L.A. (1976) The effects of climate and nutrient supplementation on the fertility of heifers in north Queensland. *Proceedings of the Australian Society for Animal Production* 11:249–52.

Slatyer, R.O. (1954) A note on the available moisture range of some northern soils. *Journal of the Australian Institute of Agricultural Science* 20:46–7.

Tothill, J.C., Mott, J.J. and Gillard, P. (1982) Pasture weeds of the tropics and sub-tropics with special reference to Australia. In: W. Holzner and N. Numata (eds) *Biology and ecology of weeds.* Dr W. Junk, The Hague, pp. 403–27.

Tothill, J.C., Nix, H.A. Stanton, J.P. and Russell, M.J. (1985) Land use and potential of Australian savanna lands. In: J.C. Tothill and J.J. Mott (eds) *Ecology and management of the world's savannas.* Australian Academy of Science, Canberra, pp. 125–41.

Williams, J. (1991) Search for sustainability: Agriculture and its place in the natural ecosystem. *Agricultural Science* 4:32–39.

Williams, J. and Coventry, R.J. (1979) The contrasting hydrology of red and yellow earths in landscapes of low relief. In: *The hydrology of landscapes of low precipitation.* Proceedings of International Symposium, Canberra, IAHS-AISH Publications No. 128, pp. 385–96.

Williams, J. and Probert, M.E. (1983) Characterisation of the soil-climate constraints for predicting pasture production in the semi-arid tropics. In: *Soil Workshop*, Townsville September, 1983. ACIAR, Canberra.

Williams, J., Day, K.J., Isbell, R.F. and Reddy, S.J. (1985) Soils and climate. In: R.C. Muchow (ed) *Agro-research for the semi-arid tropics: North-west Australia.* The University of Queensland Press, Brisbane, pp. 31–92.

Winter, W.H. (1978) The potential for animal production in tropical Australia. *Proceedings of the Australian Society for Animal Production* 12:86–93.

Winter, W.H., Mott, J.J. and McLean, R.W. (1990) Management options for increasing the production of tropical savanna pastures: Pasture response to time and treatment. *Australian Journal of Experimental Agriculture* 40:613–22.

Woinarski, J.C.Z. and Braithwaite, R.W. (1990) Conservation foci for Australian birds and mammals. *Search* 21:90–5.

Wood, I.M. and Fukai, S. (1983) The development of stable and profitable summer cropping systems in northern Australia. In: *Proceedings of 2nd Australian Agronomy Conference, Wagga Wagga, July 1983.* Australian Society of Agronomy, Melbourne, pp. 83–100.

Young, E.A. (1988) Land use and resources: A black and white dichotomy. In: R.L. Heathcote and J.A. Mabbutt (eds) *Land, water and people: Geographical essays in Australian resource management.* Allen and Unwin, Sydney.

Young, M.D. (1983) Relationships between land tenure and arid zone land use. In: J. Messer and G. Mosley (eds) *What future for Australia's arid lands?* Australian Conservation Foundation, Melbourne.

Section 6

Overview

CHAPTER 15

PROVIDING AN ENVIRONMENTALLY SUSTAINABLE, ECONOMICALLY PROFITABLE AND SOCIALLY EQUITABLE FUTURE FOR THE WORLD'S SAVANNAS

M.D. Young and O.T. Solbrig

SUMMARY

Drawing upon the material in this book, we present an overview of the world's savannas from a social and ecological perspective in search of new policy insights. Despite the different historical roots and stages in development of the savannas, some common policy prescriptions emerge. They focus on the benefits of strengthening common-property systems, considering the natural characteristics of savanna systems, developing use strategies, devolving responsibility for land management to local communities, and removing subsidies.

Many savannas contain dual socio-economic systems which appear inequitable and sometimes cause land degradation. It appears to be more effective to treat the root cause of the existing problems, rather than the symptoms, with land reform and other similar programmes.

Opportunities for greater use of the natural advantages and characteristics of savanna ecosystems are identified.

INTRODUCTION

This chapter synthesises the policy proposals and recommendations that emerge from the case and issue studies in this book. Most studies were prepared jointly by a social scientist and an ecologist, drawing on experiences within a variety of countries and presented at the 1991 Nairobi Workshop. A number of new insights emerged and participants felt that a significant step towards the integration of the paradigms that dominate ecology and economics had been achieved. A major, seemingly obvious observation that

321

arose from the workshop was that people should be recognised as a valid part of all savanna ecosystems and that the world's savannas should be studied from that viewpoint. In the past, the human dimension has been excluded from most ecological studies of savanna management. Similarly, the ecological dimension has either been excluded from most social and economic studies of the savannas or, alternatively, treated in a very simplistic manner.

SAVANNA LANDSCAPES

The world's savannas are characterised by immense diversity in climatic regimes, physical conditions, cultural and historical roots, and economic conditions. This diversity presents problems for the manager and the politician, since it requires development of flexible and variable management strategies that can be at odds with national policies and with the interests of increasingly urbanised populations.

Savanna societies are becoming progressively marginalised in political, cultural and economic terms. Although linked to national economies and through them to world markets, many savanna people have little control over their destiny. Many development, land-use and trade policies, for example, are formulated without considering their implications for the people inhabiting the world's savannas. Yet the influence of these policies on savanna land use can be dramatic. The situation is complicated further by the relatively slow speed at which savanna ecosystems reveal their responses to policy. An income tax policy that alters the relative profitability of alternative fire regimes, for example, will gradually change species composition and then, some 10–15 years later, begin to affect animal productivity and through this the welfare of local people (Holmes and Mott, this book).

DETERMINANTS AND CONSTRAINTS OF SAVANNA LAND USE

Savannas are tropical ecosystems characterised by the presence of a continuous herbaceous layer, mostly grasses, and a discontinuous canopy of trees and shrubs. Although savannas are physiognomically similar, there are major ecological differences among them in soil fertility and rainfall. The people occupying savannas differ culturally and in their historical roots and experiences. Land use reflects both the natural environment and the human society living in an area. Thus, the response of a savanna to the same policy change can vary considerably. When people extract energy and materials to fulfil their needs for food and fibre, they inevitably modify the natural environment and the potential productivity of the land. It is critical, therefore, to understand the interactions within and between both the natural biotic and abiotic components of savanna ecosystems and the socio-economic component (Solbrig, this book). The most important characteristics to consider are rainfall seasonality, soils, fire, episodic events and cultural diversity.

Rainfall seasonality

Climatically savannas are characterised by the alternation of a rainy and a dry period. The length of the rainy season, as well as the total precipitation are major constraints to the successful development of agricultural systems in savannas. The length of the rainfall season varies between 3 and 4 months in the driest savannas, to between 8 and 9 months in the wetter ones. Absolute yearly rainfall varies between 300 and 1800 mm.

Soils

The fertility of most savanna soils is extremely low and often variable at the local level. This low fertility is a major impediment to the development of agriculture in savanna regions. Yet variability in soil type, topography and rainfall has sometimes created significant opportunities within relatively small areas for the development of complex agro-pastoral systems. Along the flood plains of the Niger, for example, variability in soil fertility enables nomadic pastoralism and cropping to co-exist in a symbiotic relationship.

The limited fertility of many savanna soils, however, means that it can be relatively easy to push some savanna ecosystems irreversibly beyond their capacity to sustain use. Soil organic matter tends to be low in most areas and is easily depleted. In a number of areas the soil surface layer is prone to surface sealing, with the consequence that less water enters the soil and hence less forage is produced. These forms of land degradation are not confined to Africa and can be aggravated by technological as well as social forces. The use of mineral and protein supplements to Australia's subtropical savannas, for example, enables ranchers to maintain livestock productivity during the dry season but is gradually depleting soil organic matter. In short, the strategy is not sustainable (Holmes and Mott, this book).

Fire

Fire is one of the most powerful management tools available to savanna land-users, but misuse can lead to changes in biomass, species composition and productivity. Frequency and timing of burning are very important. Early season burning, for example, can interact with lack of rainfall to change species composition to increase the likelihood of erosion and reduce grazing potential. Fire suppression, however, can have similar effects. In short, the timing and frequency of fire are critical for savanna management. Fire management is also important for tourism and landscape amenity values.

Episodic events

All savanna landscapes are characterised by the need to manage and cope with episodic events that constrain opportunities for development. Unlike most cropping systems, savanna landscapes are non-equilibrium or multi-state systems which tend to flow from one state to another (Solbrig, this book). Traditional nomadic pastoralism fully exploits these characteristics; for example, by moving from one area to another in sympathy with seasonal conditions. Sometimes, cropping is pursued on an opportunistic basis. Likewise, wildlife tend to move across marked climatic gradients and exploit regular episodic changes in regional productivity. Significantly, there is mounting evidence that these forms of use are more economically efficient (productive) and less ecologically damaging than the sedentary systems that characterise other landscapes. Breman and Traoré (1987), for example, have found that annual liveweight gain in nomadic and semi-nomadic cattle is twice as high as that achieved by sedentary grazing.

Cultural characteristics

Many savanna cultures have deep and powerful roots and traditions which can not be changed easily. Cattle, goats, sheep and other ungulates appear to have been first domesticated some 9000 to 11,000 years ago, and about 6000 to 8000 years ago people began to take their herds out onto the savannas (Lamprey 1983). As sheep are generally unsuited to the lowland tropics, however, most savanna herds contain only cattle, goats, camels and donkeys, in various combinations.

With few exceptions, the desire for intensive agricultural development within the world's savannas is a recent phenomenon. This, however, does not mean that savanna people do not produce or eat agricultural products.

All but the wealthiest Barabaig, for example, obtain more than half their food-energy needs from flood-plain cropping and by trading cattle products for corn and other cereals (Lane and Scoones, this book).

Beyond direct sustenance values

Well noted for their rich environmental and ecological values that help to preserve ecological diversity, some savannas generate significant tourist revenue. The development of this potential in many areas, however, is still in its infancy and sometimes hindered by political instability. Substantial opportunities to enhance wildlife-related tourism exist in Brazil, Venezuela, Botswana, Kenya, Tanzania and Australia's Cape York. In Africa and Australia game ranching also has considerable undeveloped potential.

Often vast, landscapes mix monotony with surprise. Africa's large ungulates, Brazil's wide range of woody species, and the bird life within Venezuela's hyperseasonal flooded savannas provide well-known examples. Striking scenery and a unique Aboriginal heritage are unusual features of many Australian savannas.

CURRENT ISSUES

Population pressure

In Africa and Asia, population pressure, poverty and resource degradation in nearby agricultural areas are forcing people to migrate to savanna lands and crop the most promising soils which are often also key grazing areas used by nomadic pastoralists. Valley bottoms, the edges of water courses, virgin forest and ridge soils in higher rainfall areas are particularly vulnerable to invasion. Population density in the Northern Highland areas in Tanzania, for example, has reached saturation levels and local peasants are beginning to move down onto the plains traditionally used by the Maasai and Barabaig nomads.

The implications of this form of migration for savanna people are twofold. First, the quantity of land available for existing uses and people diminishes. Second, and more importantly, encroachment is leading to the breakdown of traditional grazing systems and inducing land degradation. The reasons for this are simple. The land that is most attractive to agricultural settlers is often reserved by pastoralists to provide scarce forage for dry periods and by custom is not used for much of the year. Loss of access to these *keystone grazing areas* leads to unaccustomed forms of social stress, makes many traditional resource-husbandry practices irrelevant and forces overgrazing in the areas that remain available (Lane and Scoones, this book).

Political instability and land alienation

Political conflict characterises many savanna areas. In the horn of Africa – Ethiopia, the Sudan, Northern Somalia – civil wars and military activity are common. Land-right conflicts also occur in Southern Africa, South America and Northern Australia. Land disputes occasionally result in forcible occupation. In extreme cases, such as in the Bico do Papajaio of Brazil, private armies of thugs are used to control agricultural squatters.

Land alienation for agricultural purposes is not the only problem. In Africa, land is being placed in national parks and under other forms of conservation title without adequate compensation to traditional land-users. In some areas there is also a bias towards the introduction of schemes that privatise parts of the savanna. Botswana's Tribal Grazing Lands Policy, which aimed to prevent overgrazing by settling pastoralists with large herds onto ranches, for example, has been an abject failure. Overgrazing (the problem it was supposed to reduce) and social tension increased. When the policy was introduced it was assumed that wealthier pastoralists would move their livestock onto private leasehold ranches and give up their community grazing rights (Pearce, this book). Most, however, did not give up their traditional grazing rights and, instead, began to use both resources – exacerbating the problem that the policy was supposed to resolve.

Misplaced government optimism

Governments have subsidised agricultural projects in many of the world's savannas. Australia's Ord River dam scheme and Venezuela's peanut project in the state of Anzoategui are two examples (Holmes and Mott and, also, Silva and Moreno, this book). Similar examples can be cited for almost every one of the world's savannas. All are characterised by a failure to recognise the limited economic and ecological potential of most savannas. Governments are now less optimistic about their potential to support crop production. Structural changes in the economy of Venezuela and Brazil which put more emphasis on market forces and an open economy, make such ill-planned schemes less likely to happen in the future.

Desertification and land degradation

During the 1970s and in the early 1980s much attention was devoted to investigating the risks of what was called "desertification". Little serious credence is now given to the theory of deserts advancing and swallowing up adjacent savanna landscapes (Warren and Agnew 1988; Nelson 1988). Yet in 1983 it was estimated that 17 per cent of the world's arid, semi-arid and subhumid regions had suffered some loss of productivity (UNEP 1984). More recent studies have questioned the methodology used to derive the statistics that attempt to quantify the rate at which deserts are spreading. Nevertheless, land degradation resulting from removal of the vegetation cover remains a common serious problem throughout the world's savannas (see Box 15.1). Unfortunately good statistics on the extent of land degradation are unavailable. Data that quantifies rates of decline in productivity every, say, 5 years is virtually non-existent.

The social and economic reasons for irreversible declines in production potential are complex, but collectively they are transmitted to the landscape via increased grazing pressure, deforestation for charcoal production and firewood, and inadequate agricultural practices. Fire intensity and frequency can also decrease soil nutrient content and, by reducing species diversity, reduce dry-season production potential.

Loss of biodiversity appears to be a serious problem in many savannas but as yet it is not well documented. In areas of high domestic grazing pressure, loss of animal biodiversity is aggravated by a reduction in the number of forage plants available for wildlife. It can also be aggravated by the construction of fences designed to exclude wildlife from areas grazed by domesticated animals. For example, Botswana has constructed cordon fences across its savannas in order to meet the high European Community meat hygiene standards. This prevents wildebeest migration but also prevents cattle migration into game reserves. Thus, in some cases, cordon fences may improve rather than decrease regional biodiversity (Veenendaal and Opschoor 1986).

Lost cultural identity

The last but equally serious problem in many savanna areas is a dramatic loss of cultural identity. Outside pressures for modernisation and the introduction of new policies frequently undervalue and even ignore existing cultural systems. In some cases, State policies dictate that savanna land-users abandon their traditional way of life and become "civilised". Attempts to achieve this objective have resulted in serious ecological degradation

Land degradation is usefully defined as an irreversible change in a land resource. Losses are considered irreversible if recovery would either take more than 10 years or require the investment of substantial capital of a non-recurrent nature. Three forms of land degradation can be identified:

(1) *Loss of economic potential* to produce goods and services of direct human-use value;

(2) *Loss of ecological functions* necessary to maintain ecosystem processes; and

(3) *Loss of biodiversity* at the ecosystem, species or genetic level.

Building upon these concepts, the 1970s and 1980s saw much discussion about desertification and its varying meanings.

Reviewing the literature and noting, for example, that "non-random one-off flying visits to climatically variable arid areas which are not well known to the observer, with no follow up, and often selected for having a severe problem, are likely to be biased sources of information", Nelson (1988) concluded that:

(1) There is a need for permanent national land monitoring systems to identify emerging and difficult-to-reverse forms of degradation.

(2) Research should focus on management technology and existing socio-economic systems.

(3) Policy proposals must take into account complexity and local variability.

(4) As there appear to be no global or regional solutions to most savanna and arid land degradation problems, progress will depend upon small pilot projects, community experimentation and within-country expertise.

(5) The failure and high cost of conventional projects suggests that more progress is likely through attention to enabling incentives that promote spontaneous response across the entire community. The main policy areas are land tenure, taxation and marketing.

(6) Many successful strategies will have a strong spatial dimension and involve movement across national and ecological boundaries.

Box 15.1 Desertification – a much-confused concept. Source: Nelson (1988) and extensive social and economic disruption. Partly this pressure against traditional lifestyles is due to ideological prejudices against nomadic pastoralism and a political preference for sedentary crop-based systems.

Loss of cultural identity is a particularly acute problem within Africa. Yet there is growing evidence that many traditional pastoral systems are highly rational and much more sophisticated and productive than is realised. In famine-ridden north-west Kenya where grazing-pressure-induced droughts are common, for example, the Turkana nomads have developed highly sophisticated survival strategies, which are being broken down by well-intended development programmes. In times of forage shortage the Turkana split their families, pool their surviving livestock so that they can be managed more efficiently, and exercise reciprocal arrangements with neighbours, friends and businessmen with whom they associate on a regular basis. Government-aided boarding school enrolments also increase during these periods. In 1978 the authorised total capacity of Turkana boarding schools was 995 boarders. The actual number of boarders, however, was more than double this number at 2373 (Odegi-Awuondo 1990).

Another reason for the pressure against traditional lifestyles and resulting loss of cultural identity is that often temperate development models that are totally unsuited to tropical savanna conditions have been adopted (Odegi-Awuondo 1990). Schooling systems compatible with a nomadic lifestyle, for example, still have to be developed. At the same time, however, it must be realised that many traditional pastoralists are expanding their interests. Maasai who have acquired commercial, entrepreneurial and political skills are now being seen as Kenyans (Kituyi, this book).

Overview of policy challenges

In summary, the world's savannas present a variety of problems which are both different and similar to those found in other areas. Resolving them will be difficult and will require policies sensitive to spatial and climatic variability. Since the social systems and ecological constraints are different to those found in other areas, new cross-disciplinary paradigms that account for interactions between local communities, the natural resource base, markets and the socio-political environment are necessary. People must be seen and studied as an integral part of all savanna ecosystems. Drawing on past experience, we suggest that the greatest prospects for a successful development of the world's savannas lie with production systems that:

(1) Capitalise on the natural economic, social and ecological advantages of savanna ecosystems; and

329

(2) As suggested by Nelson (1988), focus on mechanisms that promote spontaneous change.

If this suggestion, which emerged from the workshop, is correct then substantial opportunities to improve the economic productivity of savanna landscapes and the welfare of savanna people remain. Most, however, require attention to social as well as material technology. Significant opportunities include the development of tourism and game ranching systems. Another possibility is to develop savanna plant species into new products. The fruit tree *Caryocar brasiliensis*, for example, is just being brought into commercial oil production. Gum arabic production is another area of significant potential for improvement (Barbier 1990).

It may also be possible to increase livestock productivity but this will require the astute development of innovative grazing systems that gradually reclaim degraded areas. Improved marketing and credit systems could assist such processes.

In many areas formal recognition of traditional land rights may promote spontaneous development in ways that are simultaneously efficient, equitable and preserve environmental integrity. In other areas, however, land reform may be necessary.

At an international level there is also room for adjustment. Reduction in the degree of agricultural protection that exists within the European Community and inside the United States of America could provide a strong economic incentive for improved resource management. Policy changes that reduce the dependence of savanna people on the policies of other countries could also increase development opportunities.

INSTITUTIONAL AND ADMINISTRATIVE ISSUES

Alternative-use philosophies and value systems

Many seemingly indirect policies drive the pace and direction of land-use change within the savannas. Exchange rate and tariff policies, for example, have a major influence on market opportunities and, hence, grazing intensity. Similarly, institutional and administrative arrangements that determine who makes land-use decisions and who enforces land-use obligations have a dominate influence on rates of land degradation.

In order to search for options that will improve the policies that drive savanna land use, it is useful to identify the range of institutional arrangements and value systems that operate within savannas. One approach is to recognise the policy arenas that governments can use to influence savanna land use. Another approach is to focus on land-tenure arrangements and group them into:

(1)   Common-property grazing regimes;

(2)   Private-property ranching regimes; and

(3)   Dual-property regimes, where common-property and private-property ranching regimes co-exist.

Almost all agricultural land within the savannas is farmed under private-property regimes but sometimes the right to graze uncropped land is held communally.

Political and social value systems are equally hard to classify and not uniform across the savannas. Recognising the popular cry for sustainable development, the workshop that led to this book focussed on three social objectives which appear to be pursued by most governments:

(1)   Economic efficiency;

(2)   Environmental integrity; and

(3)   Equity.

In essence the search is for policy prescriptions that achieve these "3Es". A critical point, not always appreciated, is that these objectives are interdependent and that any policy prescription must attend to each simultaneously (Young, this book).

The remainder of this chapter focuses on policy opportunities that, if taken up, would assist national, regional and local governments; international agencies; and non-government organisations to improve savanna land use and meet the needs of local communities.

Understanding existing socio-economic and cultural management systems

Because of their low productivity, high variability and seasonal diversity, many savannas are managed under complex socio-economic arrangements which have evolved over centuries. They tend to be very well adapted to all but recent changes in social, economic and political conditions. Unfortunately, however, the reasons for the customs that underpin these systems tend to be poorly understood and rarely documented. This means that it can be very difficult to predict the effects of a policy change on a community, and even harder to predict how communities will adapt to changes in other parts of a country. Simplistic assumptions about the nature and resilience of existing cultural and economic systems tend to lead to environmentally degrading, economically inefficient and inequitable

331

outcomes. One of the most common mistakes is to view pastoralists in isolation rather than part of a changing national economy.

The failure of Tanzania and Canada to appreciate the implications of cropping Barabaig grazing land, and the failure of sedentary private-property ranching in Botswana, illustrate this point. In Tanzania, experts failed to realise that the fertile *muhajega* areas, whilst only grazed during the wet season, are vital to the Barabaig. Similarly, they failed to appreciate how social customs would thwart the introduction of ranching systems. Too often, the outcome of so-called savanna development programmes has been the emergence of systems that are more prone to land degradation, less efficient, less productive and less equitable that the ones they replaced. In short, *unless a policy change is fully supported by the local community, no land-use or tenure regime should be modified without a full assessment of existing cultural and ecological constraints to development, and of all the costs and benefits of the proposed change.*

Understanding land capabilities

At the same time, there is also a need to understand the limited potential of savannas to produce products of value to people. Even in areas with high rainfall, savanna land systems tend to be less productive than equivalent temperate agricultural systems. In large part this lower productivity is due, first, to the seasonal distribution of rainfall that characterises the tropical savanna climate and, second, to the low physical and chemical soil conditions common to savanna ecosystems. Too many savannas are still shrouded in a political veil that continually seeks to extract more from these ecosystems than is physically possible (Klink et al. this book). The experience described in this book is that *failed savanna development proposals and policy changes tend to be characterised by a lack of inventory and ecological assessment.*

Given the recurring uncertainties about the development potential of savannas, in many cases, a set of effective land-use policies can only be formulated after land classification and a realistic evaluation of their capability. This is particularly important in areas where there is a high probability of land degradation from the breakdown of existing management systems, market-induced overstocking or overcultivation.

Note also that *the influence of a government policy change on the socio-economic system can be relatively slow in changing land-use practice and even slower in revealing its ecological consequences.* Indeed experience suggests that many ecological and social changes become irreversible before they are recognised.

Strengthening social technology at the local level

As stated earlier at the introduction of this chapter, many savanna land-users perceive that they are being marginalised in a political, economic and social sense. At the local level, processes of economic and social differentiation may lead to different claims over land. In Africa, for example, cultivators frequently encroach onto the pastoral commons. In Latin America, small holders often squat on large estates. In Northern Australia, Aboriginals are asserting land claims over pastoral leases. Frequently, those who live in the savannas for a long time perceive that many decisions are made without consideration of their views. International and national concerns about savanna biodiversity, for example, easily translate into new tourism, wildlife and habitat conservation policies but, in doing so, often reduce options for most traditional landusers. Similarly at the regional level, infrastructure development projects have been implemented with little regard for remnant indigenous minorities.

Given all of the above, it is difficult to be specific. One can observe, however, that *administrative systems seem to be most effective when regional and state authorities are democratically accountable and institutions are responsive to local demands.* In many, perhaps most, parts of the world's savannas, however, these conditions do not apply. Subjective assessments suggest that, whenever the administrative structure is not democratically accountable, the extent of land degradation tends to be greater and there is increased poverty amongst local people. When local people do not have sufficient political power to negotiate acceptable solutions, corruption and vote buying emerges.

One characteristic common to many savannas where degradation is a problem is an absence of formally defined land-use rights and the lack of a vision that identifies a national role and charter for local people. Such loose arrangements have worked for centuries but rely upon common interests, strong dependence on the resource that is being managed, and no possibility of intervention by people who are not part of "the community". As development opportunities open up elsewhere, however, some local people become less dependent upon the savannas. Similarly, the twentieth century concept of the right of nations to govern means that intervention by people who have little affinity with the interests of local people is a real possibility.

Thus perhaps *one of the most effective ways of protecting communities from inadvertent outside intervention is to begin by formally documenting and registering traditional rights.* Once this is done, negotiation with existing land-users becomes an essential part of the development process as they hold a quasi veto power over development proposals. Such arrangements increase the probability that a policy change will, first, be consistent with local objectives and, second, will be socially equitable. They also force governments to develop social technologies that allow local communities to keep pace with material technology. Too often, material technologies, like a new vaccine or cordon fence, have been introduced without the development of parallel programs that enable local people to adjust to the new conditions that the technology creates (Scoones et al. this book). The development of new social technologies becomes particularly important when traditional land users lose their direct dependence upon savanna resources and begin to derive income from non-savanna activities.

Ecologically sensitive administrative arrangements

Savanna ecosystems tend to be much more episodic, more variable, less predictable and slower to respond than most agricultural systems. As a result, policy rules have to be more discretionary in nature and require greater budgetary flexibility. Many modern bureaucracies, however, particularly those in Africa, are uncomfortable with variable rules that require discretionary implementation at the local level and sometimes require strategic expenditure while an opportunity remains open. Administrative mechanisms that provide for flexibility, however, increase the possibility of achieving sustainable forms of savanna land use.

Devolving responsibility

Conceptually, the greater devolution of power and responsibility to local communities is a simple and seemingly obvious recommendation. Turning it into practice, however, is quite difficult because it requires careful research to identify the most appropriate way to define a community and to distribute power and responsibility within it. Case studies across the savannas suggest that *cogent arguments exist for the greater devolution of power to local people*. This applies equally to private-property and common-property regimes. Advocates for such arrangements often argue for a bottom-up rather than top-down approach, but this prescription is too simplistic. Many decisions are best taken at the household or local community level, some at a regional level, a considerable number at the national level and a few at an international level. The challenge is to find the most appropriate mix of decision-making powers. Sometimes, local communities have a more holistic appreciation of the issues and, hence, are less likely than a government department to make severe mistakes. Thus, it can be very useful to structure regional decision-making bodies so that local people have a majority vote.

334

Questions about the most appropriate level of support from higher levels of government also need to be resolved. Attention to enforcement powers is critical. Many national governments feel uncomfortable about local communities taking the law into their own hands but, while they wish to, are unable to offer an effective substitute with the consequence that well-managed common-property lands become degraded open-access lands. As indicated earlier, a "tragedy of the commons" is created by government intervention.

International aid

One of the necessary conditions for sustainable development is that there should be a significant transfer of wealth to the third world and that ways must be found to reduce the debt burden in these countries. There is no lack of options; what is missing is a collective will on the part of the richer countries to deal decisively with the issue (World Bank 1989).

With regard to aid, it has been suggested that donors should place their efforts in a longer term framework and recognise the need to strengthen institutions and develop capabilities to cope with adjustment processes that will take decades to achieve. There has also been a suggestion that donors should shift their focus from a project-by-project approach to arrangements that finance a "time-slice" of a sectoral or subsectoral programme (World Bank 1989). An alternative approach, more consistent with the above observations, would emphasise community needs for regional rather than sectoral development.

RESOURCE RIGHTS

Establishing legal frameworks for land use

In the world's tropical savannas, the diversity of land-tenure and land-right systems is, in part, a logical response to the parallel diversity in modes of land use, in stages of economic development and in population pressure but is equally strongly influenced by history, the structure of society and the distribution of power. Generally, land-right and land-tenure arrangements in many pastoral savannas still fall well short of the degree of security of tenure and control awarded to those directly engaged in many other forms of resource use. While this continues, the goals of efficient and ecologically sustainable use will not be capable of realisation. Thus, and as recognised earlier, *the elaboration of clear and enforceable rights to land is an important step towards the emergence of sustainable forms of land use*

Semi-arid grazing land in northern South Australia is leased to pastoralists under a system designed to ensure that the land is used without degradation. Resource security is made conditional upon environmental security. From 1990, rights to graze the leased land will be reissued every 14 years for a further 42 years in a roll-over framework. Each lease contains a series of covenants and conditions that require the land to be used on a sustainable basis. Legislation guarantees that any lessee who has complied with the conditions attached to a lease must be offered a new lease that runs for another 42 years. This right to the continuing reissue of a lease extends in perpetuity but every 14 years lease conditions may be changed to account for new ecological information and changes in technology. Ultimately, the roll-over process will be synchronised within regions so that monitoring becomes more efficient. Regional reviews will then be synchronised so that one region is always under review and politicians and bureaucrats become accustomed to the process and the industry's financial security is not threatened.

Lessees who have not complied with conditions may not have the term of their lease extended for another 42 years until they have repaired any damage. Once a lease condition has been breached, the onus is on the lessee to prove that the breach has been rectified and then apply for a lease extension. Between review periods, administrators have the power to order areas to be destocked, to require lessees to prepare management plans, to fine lessees for covenant breaches and, in severe cases, to cancel the lease. Lessees are required to maintain monitoring points that track vegetation changes.

The costs of this system are recovered through a rental system that fluctuates with livestock numbers and is automatically (i.e. apolitically) adjusted for inflation.

Box 15.2   Pastoral land-tenure arrangements in South Australia. Source: Young (1992)

across all the world's savannas. Such rights can be in the form of a non-transferable title held by a community, or an enforceable lease, rather than an exclusive land title, which passes onto the title-holder's heirs and successors in perpetuity (Holmes and Mott, this book; Bromley and Cernea1989). Outside the savannas, in South Australia's semi-arid areas, secure grazing rights are now being made conditional upon compliance with lease conditions (Box 15.2). Significantly, these leases make sustainable forms of land use a matter of self-interest.

Note also that it is important to take great care in defining which people are involved in and which are excluded from a body empowered to make a final decision about permitted forms of land use and the obligations that go with that permission. Generally, the policy conclusion is that *legal frameworks should be enabling, rather than prescriptive*. Under such frameworks, complementary forms of common-property and private-property regimes can co-exist (Holmes, this book and Young 1992).

Facilitating land reform

To a greater extent than in any other biome, the tenure systems that exist within many savannas tend to be a symbol of conflict and pressure for change. In many parts of Latin America and Southern Africa, for example, there are extreme inequities in land distribution. Some of these are caused by economic considerations and others by racial discrimination. Views differ about the extent of these injustices and the most appropriate way to rectify them. Options include a variety of land-redistribution programmes and attention to the policies that caused these inequities.

In many savanna systems where land redistribution has been implemented, the outcome has been a situation that is less equitable than before the reform began. A major impediment is the lack of cohesive political power among the rural dispossessed, and their increasing irrelevance to the processes of economic and political change (Holmes, this book). In Latin America, for example, the experience of the last four decades suggests that entrenched power relations tend to prevent effective land reform, either through democratic processes or through revolutionary social upheaval. In Venezuela, for example, one land redistribution programme resulted in a less rather than a more equitable distribution of land (Silva and Moreno, this book).

Increasingly there is recognition that most compulsory land-redistribution programmes fail because they treat the symptoms but not the causes of an unjust socio-economic system. Developing these arguments and applying them to South Africa, for example, Mentis and Seijas (this book) argue that reform in their country should begin with the repeal of the Native Trust and Land Act and possibly the introduction of a land tax which, acting across all land, would encourage market-driven land reform. At a more general level, Young (this book) argues for careful attention to exchange rate, monetary and tariff policies. High real interest rates, for example, can make it prohibitively expensive to reclaim a degraded savanna and highly profitable to degrade land that is in near-pristine condition.

Empowering disadvantaged minority groups

Throughout Africa and India indigenous people are being forced back into the least accessible and least productive savanna lands. Gradually submerged by dominant groups, their cultural survival is now closely linked to the acquisition of a secure title to the land they use. The title can be private or, alternatively, held in common but, if indigenous communities are to survive, whatever title is given security appears to be a necessary precondition.

In Australia, Aboriginal land rights have recently received considerable national attention with mixed outcomes. Those members of a community with a traditional interest in an area of land can apply for a non-transferable title to that land, which can not be sold or mortgaged to third parties. The Aboriginal Freehold title, as it is known, vests all land-use rights in traditional owners and provides that the title may not be transferred to other people. Those who oppose it point to the dual socio-economic system that this sets up. Others, however, see it as a necessary transitionary process that should only be changed when large numbers of Aboriginal people demand an alternative arrangement (Holmes and Mott, this book). In Latin America and some parts of Africa, mechanisms to maintain the land base of indigenous cultural groups in a way that will ensure cultural survival have yet to be developed.

Re-evaluating the merits of land privatisation

In Africa and India there is a need to further examine the merits of common-property tenure systems and compare them with privatisation options. Case studies presented at the workshop suggest that land privatisation increases the rate of land degradation and reduces productivity. Too often, those who are displaced find that they receive no compensation and, as a result, suffer increased poverty. Displaced from their traditional land, people lose the opportunity to seek and give assistance to each other without having to resort to the indignity of asking for charity. Access to land also greatly improves diet diversity and thereby the health of local inhabitants (Whittaker and Archer 1985).

The alternative approach is to strengthen existing common-property arrangements by devolving responsibility back to local communities and build upon the natural advantages of savanna systems. Sometimes this will require governments to remove or change existing arrangements. In India, for example, Gadgil argues that the way to stop undisciplined harvests of state-owned forests is to return them to local control and stop subsidising

338

Savannas secondarily derived from woodland ecosystems cover a large part of India. They perform an important service in meeting a variety of the subsistence biomass needs of a significant proportion of the rural population. Largely controlled by the State, the savannas are currently being degraded under the pressure of unregulated harvests by the commercial as well as the subsistence sector.

These undisciplined harvests are promoted by the open-access regime prevalent on State-controlled lands and by the high level of subsidies with which forest produce is supplied to the commercial sector. The withdrawal of opportunities for commercial harvest from these lands could alleviate the situation. Public lands could then be devoted to fulfilling the subsistence biomass needs of the rural population and commercial demands for non-wood forest produce, under management regimes involving the active participation of local communities.

Box 15.3   Community practice across India's savannas. Source: Gadgil (this book)

commercial logging (Box 15.3). Zimbabwe's innovative Communal Areas Management Program for Indigenous Resources (CAMPFIRE) programme is based upon this approach (Murphree and Cumming, this book; Martin, 1986).

Common-property systems

In areas where common-property management has been the norm, the consensus that appears to be emerging is one that begins by formalising the existing land-use system, and then acts to strengthen the capacity of local groups to control the rate at which these systems evolve and adapt into new ones. The definition of legal rights to land or natural resources needs to focus on key resource areas or definable property, such as trees or water sources. In some situations codified, well-regulated and functioning common-property management systems may exist. In other settings, a more flexible, opportunistic regime, underpinned by an non-transferable community title that protects those associated with it, may be more appropriate. This may be less identifiable as a "management system", but nevertheless it generally acts to regulate and manage the common-property resource. Where they operate, it is still important to codify and formalise the existing system so that local people perceive that they are responsible (solely but collectively) for the management of that resource.

Many local management systems are based on flexible, negotiated contracts, both implicit and explicit, which may be changed through negotiation, if circumstances require. The replacement of such co-ordination with rigid laws can be disruptive. This is particularly likely to be the case whenever the implementation of the formal system transfers responsibility for enforcement from the local community to a government department.

In areas where management is weak or ineffective, the re-creation and affirmation of local institutions may be a necessary precondition to the return to forms of resource management that are in the collective interest of the people affected by current practices. In taking this path, however, care must be taken to ensure that the people involved have a strong interest and dependence upon the resource in question. In communities where objectives diverge, special arrangements like a non-transferable title may be necessary to retain equity and ensure effective management.

PROVIDING INCENTIVES FOR SUSTAINABLE LAND MANAGEMENT

In the past, many policies designed to encourage sustainable savanna land use have failed because they are unprofitable or, as indicated earlier, inconsistent with social norms. From an economic perspective problems come from two sources: first, the general tendency of subsidies to mask the capacity of markets to inform land-users about the real ecological and economic capacity of the ecosystems they manage and, second, the marginal status of many savannas, which means that national economic decisions are taken for reasons that have nothing to do with the conditions that prevail within a savanna.

Recognition of the natural potential of savanna ecosystems

Almost all the case studies presented to the workshop contained examples of the adverse effects of subsidies on prospects for sustainable savanna land use. In India, for example, local communities are unable to compete with the subsidised commercial forestry sector and, hence, treat publicly owned forests as an open-access resource. Similarly, in Brazil, crop production subsidies have been paid in proportion to crop area, with the result that the area cropped is larger and the biodiversity values less than they otherwise would be. The situation is worsened by credit concession arrangements that provide a strong incentive for land owners to plough previously uncropped land rather than develop rotational-farming systems. Often subsidies mask important economic and ecological information and provide an economic incentive for savanna users to assume that governments will help them to cope with adverse climatic and economic variations. In Australia, for example, drought subsidies have tended to encourage ranchers to overstock and not to plan for periodic droughts. Recognising this, recently the government has removed most drought subsidies and has devolved responsibility for resource management to local communities and land-care groups (Holmes and Mott, this book).

Generally and with the exception of arrangements likely to speed transition or make it more equitable, subsidies tend to discourage the adoption of sustainable management strategies. Moreover, because even transitional subsidies distort information about the natural capacity of an ecosystem, they tend to be least damaging when they are introduced with an unqualified sunset clause. Experience in other countries tends to reinforce these observations. Almost all attempts to subsidise savanna land use have not succeeded in bringing about a transition to sustainable forms of savanna development. Often the cost to tax-payers and consumers has been high. Despite 30 years of subsidies and infrastructure development, those who are paid to grow crops in Venezuela's savannas, for example, are still unable to compete with imports (Silva and Moreno, this book).

Macro-economic policy

Many national and international policies also distort the information received by savanna people and encourage unsustainable forms of land use. A high real interest rate, for example, reduces the apparent opportunity cost of resource degradation. Note that currency overvaluation and restrictions on foreign currency dealings can have similar adverse effects on opportunities for economic development and induce degradation (Young, this book). As experience in Brazil illustrates, often such adverse effects are compounded by taxation policies that give preference to people who make profits in other areas. When this occurs local communities, who need to make a pre-tax as well as a post-tax profit, are disadvantaged. In Brazil these adverse policy impacts have been compounded by an inflation rate, which has averaged 40 per cent per annum, that makes savanna land an excellent hedge against inflation, and makes it even more difficult for local people to compete with those who have access to money earned in other areas (Klink et al. this book).

Trade policy

Throughout the savannas trade policies seem to have effects that are similar to those that characterise subsidies. In short, they distort information and reduce opportunity. In many industrialised countries, agricultural products are highly subsidised and trade restrictions used to protect farmers from having to compete with savanna land-users, with the result that prices are lower than they otherwise would be and opportunities for savanna development are reduced significantly. An industrialised country-led, multi-lateral reduction in tariff barriers would change this situation and, significantly, enable savanna land-users to diversify (Young, this book).

Tariff policies can also be used by countries to restrict imports and transfer wealth from rural to urban populations. In Zimbabwe, for example, import restrictions and pricing policies effectively tax savanna landusers in order to subsidise the manufacturing industry and city dwellers. Given this, and as Murphree and Cumming (this book) note, it is not surprising that small-scale commercial and peasant farmers (other than maize growers in some regions) have not prospered.

Note that tariff barriers and price support within industrialised countries have also created opportunities for the European Community to severely distort production. Case studies from Botswana and Zimbabwe both indicate that the immense profits that a few can make under the preferential trading agreements like the Lomé agreement tend to reduce prospects for a transition to sustainable savanna land use (Pearce and, also, Murphree and Cumming, this book). In Botswana, for example, in order to meet herd health standards set by the European community, large veterinary fences have been built across much of the country. This has enabled a few local ranchers to profit from access to European Community markets, but the cost has been the exclusion of wildlife and opportunities to develop game ranching as wildebeest, gazelle and other migrating species have less opportunity to pass through much of the country. Although various estimates of the returns from subsidised wildlife ranching are significantly greater than those achievable from unsubsidised cattle ranching, the decision to promote beef production has been taken (Pearce 1990).

As described by Lane and Scoones (this book), Canadian aid coupled with input subsidies supplied by the Tanzanian government have caused cropping to expand onto *muhajega* land which plays a keystone role in the grazing system maintained by the Barabaig pastoralists. The Barabaig do not agree that a national desire for self-sufficiency in wheat production justifies the destruction, without compensation, of their grazing system.

CONCLUDING COMMENTS

This chapter has attempted to draw together the observations and policy insights that are common to the world's savannas. Many of the social, ecological and economic problems found in these areas have common roots. Often at the periphery of national policy considerations, savanna people have suffered the side effects that come from activity in other areas and other sectors. Too often people who usually do not live in the savannas have had unreal expectations about their ecological and economic potential.

Recognising these characteristics, the most general recommendations that can be made suggest the following needs:

(1) To begin with a careful assessment of the ecological capacity of savanna systems;

(2) To avoid subsidies;

(3) To devolve responsibility for management to local communities; and

(4) To provide these communities with opportunities to make a significant but sustainable contribution to their economy.

REFERENCES

Barbier, E.B. (1990) The economics of controlling degradation: Rehabilitating gum arabic systems in Sudan. London Environmental Economics Centre Paper 90–03.

Breman, H. and Traoré, N. (eds) (1987) Analyse des conditions de l'élevage et propositions de politiques et de programmes. République du Mali. OECD, Paris, SAHEL D. 87:302–42.

Bromley, D.W. and Cernea, M. (1989) Management of common property natural resources: Overview of Bank experience. In: L.R.Meyers (ed) *Innovation in resource management: Proceedings of the Ninth Agriculture Sector Symposium*. World Bank, Washington DC.

Lamprey, H.F. (1983) Pastoralism yesterday and today: The overgrazing problem. In: F.Bourlière (ed) *Tropical savannas*. Elsevier, Amsterdam.

Martin, R. B. (1986) *Communal Areas Management Program for Indigenous Resources (CAMPFIRE)*. Department of National Parks and Wildlife Management, Harare.

Nelson, R. (1988) Dryland management: The "desertification" problem. World Bank Environment Department Working Paper No. 8. World Bank, Washington DC.

Odegi-Awuondo, C. (1990) *Life in the balance: Ecological sociology of Turkana nomads*. ACTS Press, Nairobi.

Pearce, D. (1990) Sustainable development in Botswana. In: D. Pearce, E. Barbier and A. Markandya. *Sustainable development: Economics and environment in the third world*. Edward Elgar, Aldershot.

UNEP (1984) General assessment of progress in the implementation of the Plan of Action to combat desertification 1978–1984. United Nations Environment Policy, Nairobi.

Veenendaal, E.M. and Opschoor, J.B. (1986) Botswana's beef exports to the EEC: Economic development at the expense of a deteriorating environment. Unpublished paper, Institute for Environmental Studies, Amsterdam.

Warren, A. and Agnew, C. (1988) An assessment of desertification and land degradation in arid and semi-arid areas. *Drylands Paper No.2*, International Institute for Environment and Development, London.

Whittaker, D.E. and Archer, F.M. (1985) Access to health care in Nourivier, Namaqualand. *South African Medical Journal* 68:663–8. Quoted in Archer, Hoffman and Danckwerts (1989).

World Bank (1989) *Sub-Saharan Africa: From crisis to sustainable growth.* Washington DC.

Young, M.D. (1992) *Sustainable investment and resource use: Equity, environmental integrity and economic efficiency.* Parthenon Press, Carnforth and Unesco, Paris.

INDEX

345

347